Chemometrics and Chemoinformatics

ACS SYMPOSIUM SERIES **894**

Chemometrics and Chemoinformatics

Barry K. Lavine, Editor
Oklahoma State University

Sponsored by the
ACS Division of Computers in Chemistry

American Chemical Society, Washington, DC

Library of Congress Cataloging-in-Publication Data

Chemometrics and Chemoinformatics / Barry K. Lavine, editor.

p. cm.—(ACS symposium series ; 894)

Includes bibliographical references and index.

ISBN–13 978–0–8412–3858–9

1. Chemometrics. 2. Chemoinformatics.

I. Lavine, Barry K., 1955- II. American Chemical Society. III. Series.

QD75.4.C45C484   2005
543'.072—dc22   2005045503

Copyright © 2005 American Chemical Society

Distributed by Oxford University Press

All Rights Reserved. Reprographic copying beyond that permitted by Sections 107 or 108 of the U.S. Copyright Act is allowed for internal use only, provided that a per-chapter fee of $30.00 plus $0.75 per page is paid to the Copyright Clearance Center, Inc., 222 Rosewood Drive, Danvers, MA 01923, USA. Republication or reproduction for sale of pages in this book is permitted only under license from ACS. Direct these and other permission requests to ACS Copyright Office, Publications Division, 1155 16th Street, N.W., Washington, DC 20036.

The citation of trade names and/or names of manufacturers in this publication is not to be construed as an endorsement or as approval by ACS of the commercial products or services referenced herein; nor should the mere reference herein to any drawing, specification, chemical process, or other data be regarded as a license or as a conveyance of any right or permission to the holder, reader, or any other person or corporation, to manufacture, reproduce, use, or sell any patented invention or copyrighted work that may in any way be related thereto. Registered names, trademarks, etc., used in this publication, even without specific indication thereof, are not to be considered unprotected by law.

PRINTED IN THE UNITED STATES OF AMERICA

Foreword

The ACS Symposium Series was first published in 1974 to provide a mechanism for publishing symposia quickly in book form. The purpose of the series is to publish timely, comprehensive books developed from ACS sponsored symposia based on current scientific research. Occasionally, books are developed from symposia sponsored by other organizations when the topic is of keen interest to the chemistry audience.

Before agreeing to publish a book, the proposed table of contents is reviewed for appropriate and comprehensive coverage and for interest to the audience. Some papers may be excluded to better focus the book; others may be added to provide comprehensiveness. When appropriate, overview or introductory chapters are added. Drafts of chapters are peer-reviewed prior to final acceptance or rejection, and manuscripts are prepared in camera-ready format.

As a rule, only original research papers and original review papers are included in the volumes. Verbatim reproductions of previously published papers are not accepted.

ACS Books Department

Contents

Preface ix

1. Chemometrics: Past, Present, and Future 1
   Barry K. Lavine and Jerome Workman, Jr.

2. Improving the Robustness of Multivariate Calibrations 15
   Steven D. Brown, HuWei Tan, and Robert Feudale

3. Interpretation and Validation of PLS Models
   for Microarray Data 31
   Fredrik Pettersson and Anders Berglund

4. Chemoinformatics: Perspectives and Challenges 41
   Ling Xue, Florence L. Stahura, and Jürgen Bajorath

5. Mathematics as a Basis for Chemistry 55
   G. W. A. Milne

6. On the Magnitudes of Coefficient Values in the Calculation
   of Chemical Similarity and Dissimilarity 77
   John D. Holliday, Naomie Salim, and Peter Willett

7. Cheminformatics and Comparative Quantitative
   Structure–Activity Relationship 97
   Rajni Garg

8. Prediction of Protein Retention Times in Anion-Exchange
   Chromatography Systems Using Support Vector Regression 111
   Curt M. Breneman, Minghu Song, Jinbo Bi, N. Sukumar,
   Kristin P. Bennett, Steven Cramer, and N. Tugcu

9. **Analysis of Odor Structure Relationships Using Electronic Van Der Waals Surface Property Descriptors and Genetic Algorithms** 127
   Barry K. Lavine, Charles E. Davidson, Curt Breneman, and William Katt

10. **Optimization of MDL Substructure Search Keys for the Prediction of Activity and Toxicity** 145
    Douglas R. Henry and Joseph L. Durant, Jr.

11. **Clustering Compound Data: Asymmetric Clustering of Chemical Datasets** 157
    Norah E. MacCuish and John D. MacCuish

12. **From Decision Tree to Heterogeneous Decision Forest: A Novel Chemometrics Approach for Structure–Activity Relationship Modeling** 173
    Weida Tong, Huixiao Hong, Hong Fang, Qian Xie, Roger Perkins, and John D. Walker

Indexes

Author Index 189

Subject Index 191

Preface

Chemometrics is an approach to analytical chemistry based on the idea of indirect observation. Measurements related to the chemical composition of a substance are taken and the value of a property of interest is inferred from them through some mathematical relation. From this definition, the message to the scientific community is that chemometrics is a process. Measurements are made, data are collected, and information is obtained with the information periodically assessed to acquire actual knowledge.

Chemoinformatics, which is a subfield of chemometrics, encompasses the analysis, visualization, and use of chemical structural information as a surrogate variable for other data or information. The boundaries of chemoinformatics have not yet been defined. Only recently has this term been coined. Chemoinformatics takes advantage of techniques from many disciplines such as molecular modeling, chemical information, and computational chemistry. The reason for the interest in chemoinformatics is the development of experimental techniques such as combinatorial chemistry and high-throughput screening, which require a chemist to analyze unprecedented volumes of data. Access to appropriate algorithms is crucial if such experimental techniques are to be effectively exploited for discovery.

Most chemists want to take advantage of chemometric and chemoinformatic methods, but many scientists lack the knowledge required to decide which techniques are the most appropriate as well as an understanding of the underlying advantages and disadvantages of each technique. The symposium entitled, *Chemometrics and Chemoinformatics*" which was held at the 224th American Chemical Society (ACS) National Meeting in Boston, Massachusetts on August 21–22, 2004, was intended to address these issues. Several research areas were highlighted in this symposium, including chemical structure representation, descriptor selection, and structure–activity correlations in large datasets.

Many applications in computer-aided drug design (e.g., diversity analysis, library design, and virtual screening) depend on the representation of molecules by descriptors that capture their structural characteristics and properties. The integration of similarity and diversity analysis with other methods is directly tied to the development of better and more realistic descriptors. The chapters in this text by Breneman and co-workers, Willet, and Henryand Durant present some recent developments in this area.

Hundreds of molecular descriptors have been reported in the literature, ranging from simple bulk properties to elaborate three-dimensional formulations and complex molecular fingerprints, which sometimes consist of thousands of bit positions. The development of new types of feature selection algorithms and the validation of such algorithms is crucial to progress in computer-aided molecular design. Because of the larger data sets, an assortment of computational methods also needs to be developed to cope with the enormous amounts of data generated. The crucial role played by computational methods in chemoinformatics is enumerated in a chapter by Bajorath and co-workers, and new algorithms for analyzing and mining structure–activity data are described in chapters by Lavine et al., MacCuish, Tong et al., and Garg.

Bioinformatics was also represented in the symposium (see the chapter by Bergund et al.). The processes of target identification, target validation, and assay development in high throughput screening need to be tied together, which is why data visualization, interpretation, and mining techniques have become so important in bioinformatics. The boundary between chemoinformatics and bioinformatics is blurring due to the development of better algorithms to analyze and visualize gene expression data and to integrate them with other information. This is critical in making such data more amenable to interpretation. By bringing together scientists from academia and industry in the United States and Europe at this symposium, we hoped to develop collaborations that could eventually lead to bridging the gap between the different types of data generated in drug discovery studies, thereby changing the whole concept of bio- and chemoinformatics.

Tutorial chapters are included in this volume. They can assist the reader by providing crucial background information. These chapters are also suitable for use in both graduate and undergraduate courses. An overview of chemometrics and principal component analysis is presented in the first chapter. Brown et al.discuss multivariate calibration and the problems that arise when using it, such as calibration transfer. The impact of mathematical and computational methods in chemistry is considered by Milne, who concludes that mathematical and computational methods have generally served only as a tool in chemistry, with the exception of graph theory, which has provided new and sometimes superior ways in which chemical structure can be viewed.

There was strong interest in this symposium among chemists in a number of different disciplines. The ACS Divisions of Analytical Chemistry and Agricultural and Food Chemistry, Inc. cosponsored this symposium in conjunction with the Division of Computers in Chemistry. Currently, there are few textbooks published on this subject, and this proceedings volume will contribute significantly to furthering the education of chemists and other scientists in this field.

Barry K. Lavine
Department of Chemistry
Oklahoma State University
455 Physical Science II
Stillwater, OK 74078–3071
(405) 744–5945 (telephone)
bklab@chem.okstate.edu (email)

Chapter 1

Chemometrics: Past, Present, and Future

Barry K. Lavine[1,3] and Jerome Workman, Jr.[2]

[1]Department of Chemistry, Clarkson University, Potsdam, NY 13699-5810
[2]Argose Incorporated, 230 Second Avenue, Waltham, MA 02451
[3]Current address: Department of Chemistry, Oklahoma State University, 455 Physical Science II, Stillwater, OK 74078-3071

Chemometrics, which offers the promise of faster, cheaper, and better information with known integrity, has enjoyed tremendous success in the areas related to calibration of spectrometers and spectroscopic measurements. However, chemometrics has the potential to revolutionize the very intellectual roots of problem solving. A chemometric based approach to scientific problem solving attempts to explore the implications of data so that hypotheses, i.e., models of the data, are developed with greater awareness of reality. It can be summarized as follows: (1) measure the phenomena or process using chemical instrumentation that generates data rapidly and inexpensively, (2) analyze the multivariate data, (3) iterate if necessary, (4) create and test the model, and (5) develop fundamental multivariate understanding of the process. This approach does not involve a thought ritual; rather it is a method involving many inexpensive measurements, possibly a few simulations, and chemometric analysis. It constitutes a true paradigm shift since multiple experimentation and chemometrics are used as a vehicle to examine the world from a multivariate perspective. Mathematics is not used for modeling per se but more for discovery and is thus a data microscope to sort, probe, and to look for hidden relationships in data.

© 2005 American Chemical Society

1. Introduction

Chemometrics is an approach to analytical and measurement science based on the idea of indirect observation[1]. Measurements related to the chemical composition of a substance are taken, and the value of a property of interest is inferred from them through some mathematical relation. Chemometrics works because the properties of many substances are uniquely defined by their chemical composition.

The actual term chemometrics was first coined in 1975 by Bruce Kowalski in a letter that was published in the Journal of Chemical Information and Computer Science[2]. Motivation for the development of this subfield of analytical chemistry was simple enough at the time. The dramatic increase in the number and sophistication of chemical instruments triggered interest in the development of new data analysis techniques that could extract information from large arrays of chemical data that were routinely being generated. Much of the growth and interest in the field of chemometrics that continues to occur is driven by the press of too much data.

Several definitions for chemometrics have appeared in the chemical literature since the inception of the field[3]. The message communicated by these definitions to the industrial and academic community is that chemometrics is a process. Measurements are made, and data are collected. Information is obtained with the information periodically assessed to acquire actual knowledge. This, in turn, has led to a new approach for solving scientific problems: [1] measure a phenomenon or process using chemical instrumentation that generates data swiftly and inexpensively, [2] analyze the multivariate data, [3] iterate if necessary, [4] create and test the model, and [5] develop fundamental multivariate understanding of the measured process. The chemometrics approach does not involve a thought ritual. Rather, it is a method involving many inexpensive measurements, possible a few simulations, and chemometric analysis. It constitutes a true paradigm shift since multiple experimentation and chemometrics are used as a vehicle to examine the world from a multivariate perspective. Mathematics is not used for modeling per se but more for discovery and is thus a data microscope to sort, probe, and to look for hidden relationships in data.

The chemometrics approach explores the implications of data so that hypotheses, i.e., models of the data, are developed with a greater awareness of reality. This exploratory data mining approach is in some respects more rigorous than simply formulating a hypothesis from a set of observations, since a variety of techniques can be used to validate the model with predictive success being the most powerful. If predictive success can be repeated and found to hold well for the observed experimental data, then the successful model can be articulated into cause and effect, not just simple correlation.

This new paradigm for learning can be summarized as follows: fail first, fail cheap and move on. Failures provide the necessary learning for future successes. The size of your scrapheap indicates the learning potential for future successes. This new approach, which looks at all of the data using multivariate methods, is the basis of combinatorial chemistry, which has revolutionized drug discovery [4]. Although the pharmaceutical industry has embraced many aspects of this approach, few chemists in academia actually take advantage of it. Chemometrics is considered to be too complex, and the mathematics can be misinterpreted. Problems arise in the implementation and maintenance of these methods. There is a lack of official practices and methods associated with chemometrics. This, despite the clear advantages of chemometrics, which include but are not limited to: speed in obtaining real-time information from data, extraction of high-quality information from less well resolved data, clear information resolution and discrimination power when applied to higher order data, and improved knowledge of existing processes.

2. Soft Modeling in Latent Variables

The focus of chemometrics has been the development of correlative relationships using several (measurement) variables for prediction of physical or chemical properties of materials. These correlative relationships are based on sound chemical principles, which is why they are so robust. A unique methodology known as soft modeling in latent variables[5] is used to forge these relationships. Soft modeling is based on a simple premise, that signal is the part of the data that describes the property or effect of interest, and noise is everything else. Noise can be systematic (e.g., sloping base line) or random (e.g., white noise). Because our observations are the sum of both parts, the signal is often hidden in the data. Using methods based on variance, multivariate data can be separated into signal and noise.

Consider a set of 100 point spectra of gasolines, each with a different octane number. If the correct spectroscopy has been performed, most of the variation observed in these spectra should be about changes in octane number, not random or instrumental error. The variation in these 100 point spectra can be depicted by the scatter of the gasoline samples in a data space whose coordinate axes are the absorbances of the wavelengths constituting the spectra. Scatter of points in this space is a direct measure of the data's variance. Octane number would be expected to be the primary contributor to the scatter. Figure 1 shows a plot of ten gasoline samples in a data space whose coordinate axes are the absorbance values at 3 wavelengths. A line defining the direction of increasing octane number has been drawn through the 3-dimensional measurement space. Most of the gasoline samples can be found on this line. The distance between each data point and the line is the variance of the data that is not explained by octane number. One can conclude that most of the variance or scatter in these 3 measurements is correlated to octane number.

Multivariate analysis methods based on variance constitute a general approach to data analysis since these methods can be used to explore, calibrate, and classify data. This approach to data analysis is called soft modeling in latent variables. It is synonymous with chemometrics. Soft modeling is possible because chemical data sets are usually redundant. That is, chemical data sets are not information rich. This occurs because of the way the data are taken. Consider an absorbance spectrum or a chromatogram. One sensor channel – the absorbance at a particular wavelength or the response at a set time in a chromatogram – is often related to the next channel, which results in the desired information being obscured by redundancies in the data. Redundancy in data is due to collinearity among the measurement variables. Consider a set of samples characterized by two measurements, X_1 and X_2. Figure 2 shows a plot of these data in a 2-dimensional space. The coordinate axes or basis vectors of this measurement space are the variables X_1 and X_2. There appears to be a relationship between these two measurement variables. This relationship suggests that X_1 and X_2 are moderately correlated since fixing the value of X_1 limits the potential range of values for X_2. If the two variables, X_1 and X_2, were uncorrelated, the entire area of the enclosed rectangle would be fully populated by data points. Because information can be defined as the scatter of points in a measurement space, correlations between measurement variables decrease the information content of this space. Data points are restricted to a small region of this measurement space and may even lie in a subspace when the measurement variables are highly correlated. This is shown in Figure 3. X_3 is perfectly correlated with X_1 and X_2 since X_1 plus X_2 equals X_3, which is why the 7 data points in Figure 3 lie in a plane even though each data point has 3 measurements associated with it.

3. Principal Component Analysis

Variables that are highly correlated or have a great deal of redundancy are said to be collinear. High collinearity between variables – as measured by their correlation – is a strong indication that a new coordinate system can be found that is better at conveying the information present in the data than axes defined by the original measurement variables. This new coordinate system for displaying the data is based on variance. The principal components of the data define the variance-based axes of this new coordinate system. The first principal component is formed by determining the direction of largest variation in the original measurement space of the data and modeling it with a line fitted by

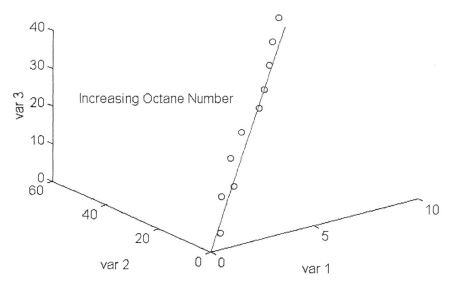

Figure 1. Ten gasoline samples plotted in a data space whose coordinate axes are the absorbance values of 3 wavelengths. A line representing the direction of increasing octane number has been drawn through this space. Most of the variation in the data is correlated to octane number.

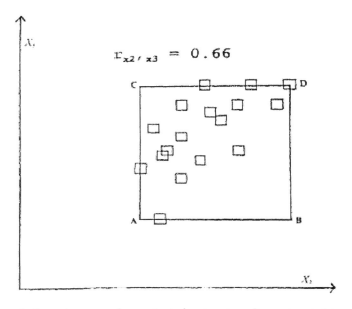

Figure 2. Seventeen samples projected onto a two-dimensional data space defined by the variables x_1 and x_2. A, B, C, and D represent the smallest and largest values of x_1 and x_2. (Adapted from NBS J. Res., 1985, 190(6), 465-476.*)*

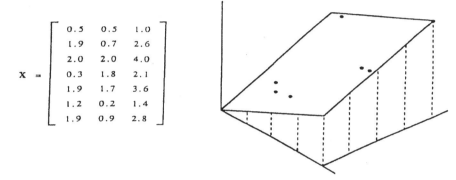

Figure 3. Seven hypothetical samples projected onto a 3-dimensional space defined by the measurement variables X_1, X_2, and X_3 (right). The coordinates of each sample point can be found in the accompanying data matrix (left). The columns of the data matrix demote the variables and the rows give the coordinates of each sample. (Adapted from Multivariate Pattern Recognition in Chemometrics, *Elsevier Science Publishers, Amsterdam, 1992.)*

linear least squares (see Figure 4). This line will pass through the center of the data. The second principal component lies in the direction of next largest variation. It passes through the center of the data and is orthogonal to the first principal component. The third principal component lies in the direction of next largest variation. It also passes through the center of the data, is orthogonal to the first and second principal component, and so forth. Each principal component describes a different source of information because each defines a different direction of scatter or variance in the data. The orthogonality constraint imposed by the mathematics of principal component analysis also ensures that each variance-based axis is independent.

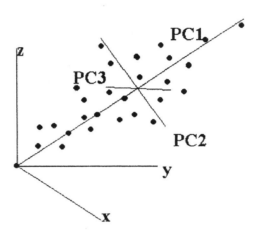

Figure 4. Principal component (PC) axes defining a new set of basis vectors for the measurement space defined by the variables X, Y, and Z. The third principal component describes only noise in the data.

One measure of the amount of information conveyed by each principal component is the variance of the data explained by the principal component, which is expressed in terms of its eignevalue. For this reason, principal components are arranged in order of decreasing eigenvalues. The most informative principal component is the first and the least informative is the last. The maximum number of principal components that can be extracted from the data is the smaller of either the number of samples or the number of variables in the data set, as this number defines the largest possible number of independent axes in our data.

If the data have been collected with due care, one would expect that only the larger principal components would convey information about the signal since most of the information in the data should be about the property or effect of interest that we seek to study. In other words, signal variation should be larger than noise variation, which means that we should discard principal components with small eigenvalues. However, the situation is not as straightforward as is implied. Each principal component describes some amount of signal and some amount of noise in the data because of accidental correlations between signal and noise. The larger principal components primarily describe signal variation, whereas the smaller principal components essentially describe noise. By discarding the smaller principal components, noise is discarded but so is a small amount of signal. However, the reduction in noise more than compensates for the biased representation of the data that occurs from discarding principal components that contain a small amount of signal, but a large amount of noise. The approach of describing a data set in terms of important and unimportant variation is known as soft modeling in latent variables.

Principal component analysis (PCA)[6-8] takes advantage of the fact that a large amount of data generated in scientific studies has a great deal of redundancy and therefore a great deal of collinearity. Because the measurement variables are correlated, 100-point spectra do not require 100 independent orthogonal axes to define the position of a sample in the measurement space. Using principal component analysis, the original measurement variables that constitute a correlated axes system can be converted into an orthogonal axes system, thereby removing correlations and forcing the axes to be independent. This requirement dramatically reduces the dimensionality of the data since only a few independent axes are needed to describe the data. The spectra of a set of samples may lie in a subspace of the original 100-dimensional measurement space and a plot of the two or three largest principal components of the data can help one to visualize the relative position of the spectra in this subspace.

By examining the principal components of the data, it is possible to identify important relationships in the data, that is, find similarities and differences among the samples in a data set, since each principal component captures a different source of information. The principal components which describe important variation in the data can be regressed against the desired property variable using linear least squares to develop a soft calibration model. Principal components that describe the property of interest are called latent variables.

With PCA, we are able to plot the data in a new coordinate system based on variance. The origin of the new coordinate system is the center of the data, and the coordinate axes of the new system are the principal components of the data which primarily describe signal. This variance based coordinate system will be different for each data set. With this new coordinate system, we can uncover the relationships present in the data. PCA is actually using the data to suggest the model, which is a new coordinate system for our data. The model is local since the model center and the principal components will be different for each data set.

The focus of PCA (and soft modeling) is signal, not noise. Furthermore, PCA based soft models are both linear and additive.

4. Partial Least Squares

Soft modeling in latent variables is central to many of the more popular methods in chemometrics. For example, a modeling method called partial least squares (PLS) has come to dominate the practice of multivariate calibration in near infrared spectroscopy because of the quality of the models produced and the ease of their implementation due to the availability of PLS software. Only a summary of the PLS method is provided here because the statistical basis of this method has been extensively discussed at many levels[9-12] including fairly introductory treatments, discussions oriented towards spectroscopists and more advanced discussions.

PLS was originally developed by Herman Wold as an alternative to classical least squares for modeling collinear data. Motivation for developing PLS was simple enough: approximate the design space of the original measurement variables with one of lower dimension. However, the eigenvectors in PLS are developed simultaneously along with the calibration model so that each eigenvector is a linear combination of the original measurement variables which has been rotated to ensure maximum correlation with the property information provided by the response variable. Because of the rotation, which attempts to find an appropriate compromise between explaining the measurement variables and predicting the response variable, surrogate variables generated by PLS are often better at capturing information relevant to the calibration than a corresponding PCA model. Confounding of the desired signal by interferents is usually less of a problem in PLS than in PCA since PLS utilizes both the response and measurement variables to iteratively determine the PLS components.

Both PLS and PCA use the technique of linear least squares, which involves an averaging of the data. Hence, both techniques will yield features that contain more information about the calibration than any of the original measurement variables provided that most of the original measurement variables contain some information about the problem. Suppose the relation between absorbance and concentration is linear for a large number of wavelengths but the sensitivity and degree of linearity at each wavelength is different. Averaging would then help. If only a few wavelengths have sufficient sensitivity and linearity, averaging of all the wavelengths would serve only to amplify the noise in the data. Furthermore, very little would be gained by averaging wavelengths that possess similar sensitivity and linearity.

5. Applications of Chemometrics

Chemometrics is an application driven field. Any discussion of this subject cannot and should not be done without focusing on so-called novel and exciting applications. Criteria used to select these application areas are based in part on the number of literature citations and in part on the perceived impact that developments in these particular areas will have on chemometrics and analytical chemistry. The four application areas that are highlighted in this book are image analysis (Chapter 3), sensors (Chapter 2), chemoinformatics (Chapters 6-14), and bioinformatics (Chapters 4 and 5).

Image analysis attempts to exploit the power gained by interfacing human perception with cameras and imaging systems that utilize the entire electromagnetic spectrum. Insight into chemical and physical phenomenon can be g arnered w here t he s uperior p attern r ecognition o f h umans o ver computers provides us with a strong argument to develop chemometric tools for imaging. These include tools for interpretation, creation or extraction of virtual images from real data, data compression and display, image enhancement, and three-dimensional views into structures and mixtures.

Chemometrics has an even greater potential to improve sensor performance than miniaturization of hardware. Fast computations combined with multivariate sensor data can provide the user with continuous feedback control information for both the sensor and process diagnostics. The sensor can literally become a self-diagnosing entity, flagging unusual data that arises from a variety of sources including a sensor malfunction, a process disruption, an unusual event, or a sampling issue. Single sensors or mixed arrays of different types of sensors can be n etworked i nto a n i ntelligent s ystem, w hich c an s end an alarm to a human operator when questions or unusual circumstances arise.

Chemoinformatics encompasses the analysis, visualization, and use of chemical information as a surrogate variable for other data or information. Contemporary applications of chemoinformatics include, among others: diversity analysis, library design, and virtual screening. Chemical structure representation through d escriptors t hat c apture t he s tructural c haracteristics a nd p roperties of molecules is an unsolved problem in chemoinformatics. Relating chemical structure to biological activity or physical and chemical properties is not a new endeavor. The ability to perform this task on a large data set, however, presents challenges that will require an assortment of new computational methods including new methods for analysis and visualization of data.

Microarrays allow the expression level of thousands of genes or proteins to be measured simultaneously. Data sets generated by these arrays consist of a small number of observations (e.g., 20-100 samples) on a very large number of variables (e.g., 10,000 genes or proteins). The observations in these data sets often have other attributes associated with them such as a class label denoting the pathology of the subject. Finding genes or proteins that are correlated to these attributes is often a difficult task since most of the variables do not contain information about the pathology and as such can mask the identity of the relevant features. The development of better algorithms to analyze and to visualize expression data and to integrate it with other information is crucial to making expression data more amenable to interpretation. We would like to be able to analyze the large arrays of data from a microarray experiment at an intermediate level using pattern recognition techniques for interpretation. At the very least, such an analysis could identify those genes worthy of further study among the thousands of genes already known.

Molecular dynamic (MD) simulations using modified versions of the conformational flooding technique when coupled with the data reduction and sorting abilities of multivariate analysis can provide critical biomolecular information on structure function relationships. Previous studies have demonstrated the need for extracting meaningful information from complex and large MD simulation data sets, while simultaneously identifying the features of the data that are responsible for the structural variation. Of particular interest is the development of methods for the reduction of large, highly dimensional information and information rich bioinformatic data sets to simple two- and three-dimensional representations enabling fundamental scientific discoveries and a dvances l eading t o a n ew g eneration a nd u nderstanding o f biochemistry. However, methodological developments should not be limited to statistical methods for mining data. Integration of computational modeling, statistics, chemometrics, and experimental evaluation will also be crucial to advances in understanding the microscopic details of biomolecules leading to a clear description of how biomolecular motion correlates with biological function. Methodology that will be developed as part of this research can provide researchers with a powerful new tool that could be applied to a wide range of problems that suffer from structural complexity including the development of synthetic agents to affect transcription, where the lack of sequence specificity prevents the utilization of existing agents to regulate a specific gene.

6. Conclusions

The field of chemometrics is in a suitable position to enter into a variety of important multivariate problem solving issues facing science and industry in the 21st century. The ever expanding endeavors of i maging, s ensor d evelopment, chemoinformatics, bioinformatics, machine learning, evolutionary computations, and multivariate data exploration will also prove to be challenging opportunities for new scientific insight and improved processes.

References

1. Lavine, B. K.; *Analytical Chemistry*, **1998**, 70, 209R-228R.
2. Kowalski, B. R.; *J. Chem. Inf. Sci.*, **1975**, 15, 201.
3. Workman, J.; *Chemolab*, **2002**, 60(1), 13-23.
4. Wold, S.; Sjostrom, M.; Andersson, P.M.; Linusson, A.; Edman, M.; Lundstedt, T.; Norden, B.; Sandberg, M. "Multivariate Design and Modeling in QSAR, Combinatorial Chemistry, and Bioinformatics" in *Proceedings of the 12th European Symposium on Structure-Activity Relationships-Molecular Modeling and Prediction of Bioactivity*, Jorgensen, K. G. (Ed.), Kluwer Academic/Plenum Press, New York, **2000**, pp. 27-45.
5. Lavine, B. K.; Brown, S. D.; *Managing the Modern Laboratory*, **1998**, 3(1), 9-14.
6. Jackson, J. E.; *A User's Guide to Principal Component Analysis*, John Wiley & Sons, New York, **1991**.
7. Jolliffe, I. T.; *Principal Component Analysis*, Springer Verlag, New York, **1986**.
8. Wold, S.; Esbensen, K.; Geladi, P.; *Chemolab*, 1987, 2(1-3), 37-52.
9. Geladi, P.; Kowalski, B.R.; *Anal Chim. Acta*, **1986**, 185, 1-17.
10. Geladi, P.; Kowalski, B. R.; *Anal Chim. Acta*, **1986**, 185, 19-32.
11. Martens, H.; Naes, T.; *Multivariate Calibration*, John Wiley & Sons, Chichester, U.K., **1989**.
12. Frank, I. E.; Friedman, J. H. *Technometrics*, **1993**, 35(2), 109-135.

Chapter 2

Improving the Robustness of Multivariate Calibrations

Steven D. Brown[1], HuWei Tan[1,2], and Robert Feudale[1]

[1]Department of Chemistry and Biochemistry, University of Delaware, Newark, DE 19716
[2]Current address: InfraredX, 125 Cambridge Park Drive, Cambridge, MA 02140

[2]InfaredX, 125 Cambridge park Drive, Cambridge, MA 02140
Multivariate calibration models are of critical importance to many analytical measurements, particularly for those based on collection of spectroscopic data. Generally, considerable effort is placed into constructing a multivariate calibration model because it is often meant to be used for extended periods of time. A problem arises, though, when the samples to be predicted are measured under conditions that differ from those used in the calibration. The changes in spectral variations that occur in these situations may be sufficient to make the model invalid for prediction under the changed conditions. Various standardization and preprocessing methods have been developed to enable a calibration model to be effectively updated to reflect such changes, often eliminating the need for a full recalibration. This chapter details recent efforts aimed at improving the robustness of multivariate calibrations to changes in the chemical system and to changes in the measurement process.

1. Introduction

Multivariate calibration is a useful tool for extracting chemical information from spectroscopic signals. It has been applied to various analytical techniques, but its importance has been manifested in near-infrared (NIR) spectroscopy. The most commonly used multivariate methods for chemical analysis are partial least squares (PLS) regression and principal component regression (PCR), where factors that relate to variation in the response measurements are regressed against the properties of interest. Ideally, each factor added to the model would describe variation relevant for predicting property values. In NIR spectra, however, the first few factors that describe majority of the spectral variation usually account for baseline shifts and various instrumental effects.

A practical limitation to multivariate calibration occurs when an existing model is applied to spectra that were measured under new sampling or new environmental conditions or on a separate instrument. Even if samples with identical amounts of analyte are measured, the spectral variation that is captured by the model will differ because of the different contributions from the sample matrix, the instrumental functions and the environment of the measurement. For this reason, a model developed on one instrument can generally not be used on spectra obtained from a second instrument to provide accurate estimates of calibrated property values. However, even the subtle changes in the matrix, the environment, sampling and instrumental function that occur over time can degrade a calibration model.

Improving a calibration model to make it less sensitive to changes in sample matrix, instrumental response function and environmental conditions is a topic that has usually been considered in connection with other goals. Most often, the subject is connected with the transfer of a multivariate calibration model. Various methods for calibration transfer exist in the literature, and most have been discussed in recent reviews (*1-3*). As these reviews discuss, the transfer of a multivariate calibration amounts to making a mathrematical correction. The degree to which the correction succeeds depends on the complexity of the effect being compensated and the needs of the user.

It is best if the whole issue of calibration model transfer can be avoided, for example by careful maintenance of instruments so that the instrumental contribution to the multivariate measurement is made as constant as possible. Wide deployment of a calibration model is easier if the target instruments and the development instrument are very similar, so it is not surprising that many users of multivariate calibration will specify a particular brand and model of instrument be used throughout for the measurements used in the calibration. Standardization of the instrumental aspects makes the task of calibration model transfer from instrument to instrument all the easier. In rare cases, it is possible to make instruments functionally identical, though the effort needed to do so is substantial *(4)*.

Like standardizing on a single brand and model of instrument, careful preprocessing of the data may also be an aid in making a multivariate calibration model useful under a wider range of conditions. In fact, some authors have observed that with proper preprocessing, multivariate calibration models can be used on multiple instruments without change *(5,6)*. It should be noted that, as with stardardizing instruments and transferring calibration models, preprocessing efforts have met with varying success when applied generically. However, there are now new ways to assist in preprocessing to help improve the quality of a calibration model. As will be shown, these methods offer some promise in improving the usability and transferability of multivariate calibrations.

The objective of this Chapter is to consider a few new preprocessing methods that this research group has developed to help improve the robustness of calibration models. Our focus is on robustness in sense of the process analytical chemistry, where the goal is to build and maintain a calibration to predict specific targets from a chemical process that varies over time. It will be assumed throughout this chapter that the signal is a spectrum measured at defined wavelengths and that the property of interest is the analyte concentration, although properties other than concentration and other types of signals also fit the discussion.

2. Using Local Analysis to Create Robust Calibration Models

There are several circumstances that can introduce variations in the new spectra that were not included in the calibration step. The presence of these un-modeled variations in the new data can lead to strongly biased predictions from multivariate calibration models. Essentially, three situations exist that can render a model invalid. The first includes changes in the physical and/or chemical constitution of the sample matrix. These changes can result from differences in viscosity, particle size, surface texture, etc., that may occur between batches. Batch differences may arise from some change in the starting materials, sample preparation, or experimental design. One can even see effects from variations in the presentation of the sample to the spectrometer. These variations are often overlooked but can occasionally be particularly difficult to discover and compensate when they occur.

The second situation arises from changes in the instrumental response function. These changes can occur when the data to be predicted are collected on an instrument different from that used to build the model.

The third situation that may render a model invalid occurs when the instrument's environment changes over time. Temperature and humidity variations can have a strong influence on measurement values by causing shifts in absorption bands and nonlinear changes in absorption intensities in the spectra.

These contributions to signal variation that changes between a calibration and its use in prediction are difficult to fully compensate, and it is often necessary (and almost always ideal) to provide a full recalibration. Because the instrumental effects noted above can occasionally be of the same magnitude as the spectral contributions of the desitred analyte, a mathematical correction cannot be expected to fully correct for all changes, a situation most spectroscopists will readily emphasize. The ASTM method for near-infrared spectral calibration (ASTM E1655-00) takes this view, not surprisingly (7). However, neither the instrument nor its environment is static, despite the best efforts of the user. Significant changes in the instrument's behavior are to be expected over time, and as a consequence any calibration (or recalibration) effort has a useful lifetime, and that lifetime unfortunately can be quite short. Further, the cost of a full recalibration can be quite high, a consequence of the number of samples and replicates needed to get adequate calibration established. When the cost of recalibration is considered in connection with the limited lifetime of a calibration, any methodology that extends the useful lifetime of a calibration is clearly worthwhile.

In some cases, the response of the new samples is not greatly affected by the new measurement conditions and the existing model can be used without applying any corrections. When this is not the case, some strategy must be used to avoid biased predictions that can result from using a flawed calibration model. If the future sources of variation can be identified, they can be incorporated into the model during calibration. The model can also be updated with new samples to account for the non-modeled variance. Standardization methods can be applied once the model is already in use to account for the new variations by transforming the spectral responses, model coefficients, or predicted values. These topics are discussed in a recent critical review of standardization methodology (*3*).

At the outset, it should be noted that there are differing interpretations of the term "robust calibration." One could refer to "robust" in the sense that the calibration could be "robust to small, *expected* variations in measurement, sampling and processing" as would be needed in a regulated industry, or we might refer to robust in the sense that is relevant to process analytical measurements, where the robustness of a calibration model refers to its resistance to the effects of *unintended and unexpected* variations in the measurements. This Chapter focuses on the latter interpretation of the term.

The way in which a process calibration model is made more "robust" - that is, resistant to the effects of unintended measurement or other variation on the performance of the calibration model - involves one of two philosophies: either make the model small, to minimize the chance that effects external to the calibration can become a part of the calibration, or make the calibration model big enough to span the effects that might arise from variations external to the main relationship.

The first philosophy involves isolating key wavelengths (if spectra are involved) and using as few as possible to build a calibration model. This local modeling approach is well-known in near infrared (NIR) spectrometry, having long been advocated by spectroscopists such as Karl Norris and others, including Howard Mark and Tomas Hirschfeld. Their approach involves isolating and using selected pieces of data to build calibration models. The goal is to find the most relevant signals for calibration while rejecting as much irrelevant signal as possible. Three ways of locating the most relevant signal involve use of wavelength selection schemes, projection methods and wavelet analysis. A robust model can be built by using variables that are either insensitive or are less sensitive to the variations in the experimental conditions. Successful application of sensor selection is dependent on finding variables that are insensitive to the offending variation yet sensitive to quantifying the analyte.

Global modeling - the second approach - has been promoted by chemometricians such as Svante Wold and Bruce Kowalski, and this is often used in calibration modeling of complex systems where finding individual wavelengths for a small model of the sort described above may be problematic. The idea here is that there are variations correlated to the desired property that can be discovered in the presence of irrelevant variations though use of chemometric methods such as partial least squares (PLS) regression. For this method to be successful, the calibration must be developed in a way that the variation that can occur in the *use* of the model must be generated and spanned in the calibration step for that model. In a chemical process, historical data, especially of those from process upsets and outliers are sometimes used to help the calibration span spectra from these (rare?) events.

While there are many ways to create calibration models from single wavelengths, ranging from the *ad hoc* approaches seen in the early NIR literature to the more systematic and highly automated approaches based on simulated annealing and genetic algorithms mentioned above, there are relatively few methods for the systematic upweighting and downweighting of spectral bands. Given the nature of NIR spectra, such an approach makes spectroscopic sense and might be one that would offer improved robustness over conventional modeling methods.

Two approaches come to mind in selecting bands from multivariate spectra:

- Select the band on the basis of a strong covariance relationship with the property (and, possibly equivalently, reject data from bands with weak or no covariance relationship with the property); and,

- Select the band on the basis of its frequency composition (shape) and its correlation to the property.

As is well-known, calibration relationships built with partial least squares regression attempt to weight the data based on the strength of the covariance of the predictor variables and the property variables. Those having weak relationships with the property will be deweighted in the PLS calibration calculations. Global modeling methods rely on the weighting by PLS to correct for changes in the variation of components not directly connected to the calibration, but the weighting process is imperfect and some extraneous variation will become a part of the calibration model, often through accidental correlation, and changes in the weakly correlated signals will degrade the calibration model relative to the performance available from a model that avoids any contributions from external effects by some sort of variable selection.

In the following discussion, matrices are denoted by capital bold characters (**X**, **Y**), column vectors by small bold characters (**p**, **t**), and row vectors by transpose vectors ($\mathbf{p}^T$). Indices are designated by small non-bold characters (j), index limits by capital non-bold characters (K), and constants by small italic characters (r).

3. Projection Methods for Improving Model Performance

One way to avoid contributions from weakly correlated variables in a calibration is to remove them prior to regression by projecting the data into a new space so that the contributions of variables uncorrelated to the property are reduced to zero. This is the principle behind orthogonal signal correction (OSC) and its close relatives (8,9), a method that has received a good deal of scrutiny of its use in simplifying the modeling of systems that change with time or with external factors.

OSC preprocessing of a data set aims at removing the largest systematic variation in the response matrix **X** that is orthogonal to the property matrix **Y** prior to multivariate calibration. Consider the equation $\mathbf{X} = \mathbf{R} + \mathbf{Z} + \mathbf{E}$ where **R** is a matrix containing the desired signal, **Z** is the matrix of undesired systematic variation, and **E** corresponds to random noise. Since **Z** can be separated from **R**, it is possible to find a projection subspace that will eliminate **Z**. If **Z** contains spectral variation that is orthogonal to the proiperty **Y**, the orthogonal compliment is estimated by $\hat{\mathbf{Z}} = \mathbf{T}_\perp \mathbf{P}_\perp^T$ where $\mathbf{P}_\perp$ is a set of basis vectors excluding the spectral contribution from **Y** and $\mathbf{T}_\perp$ is the scores matrix in a subspace spanned by the basis defined in $\mathbf{P}_\perp$. Mathematically, the basis defining $\mathbf{P}_\perp$ must be orthogonal to the target analyte **Y**, a requirement which cannot be satisfied in some cases. Usually, no more than two orthogonal components can be found in the space, due to the orthogonality constraint, but sometimes no solution can be obtained.

In order for any signal pre-processing method to improve a PLS model, prediction must be enhanced (as indicated by lower RMSEP values) or the overall model size (as measured by the OSC components + PLS latent variables) must be reduced. In many cases, pre-processing with OSC merely reduces the number of latent variables in the PLS calibration model by the number of OSC components removed, without any improvement in prediction performance. One sees similar performance from embedded OSC filtering, as found in O2-PLS, too. One reason for this occurrence is that OSC filtering removes global, systematic variance, while spectra are inherently local and multiscale in nature. Another reason is that most OSC algorithms impose a strict orthogonality constraint, and this constraint may result in an inadequate amount of weakly relevant variation being removed from the spectral data.

In recent literature, it was shown that by relaxing the orthogonality constraint, less complex models with better prediction quality can be obtained *(10,11)*.

3.1 Piecewise OSC Preprocessing

To avoid some of the limitations mentioned above, a piecewise OSC algorithm (POSC) was recently developed in this research group. Our aim was to relax the too-restrictive orthogonality constraint by processing the spectral data set in a piecewise fashion to account for local spectral features in frequency domain.

A simplified algorithm for POSC filtering is summarized here. In a given window of the spectrum, first calculate the first principal component of $\mathbf{X}$ and compute the normalized weight vector $\mathbf{w}$ from the loading vector $\mathbf{p}$ obtained from $\mathbf{w} = \mathbf{p}/(\mathbf{p}^T\mathbf{p})$. Next, orthogonalize the score vector $\mathbf{t}$ to $\mathbf{Y}$, and compute a new loading vector $\mathbf{p}^*$ by the relations $\mathbf{t}^* = \mathbf{t} - \mathbf{Y}(\mathbf{Y}^T\mathbf{Y})^{-1}\mathbf{Y}^T\mathbf{t}$ and $\mathbf{p}^* = \mathbf{X}^T\mathbf{t}^* / (\mathbf{t}^{*T}\mathbf{t}^*)$. Compute the filtered matrix $\mathbf{E}$ by subtracting the orthogonal component according to $\mathbf{E} = \mathbf{X} - \mathbf{t}^*\mathbf{p}^*T$. To remove another orthogonal component, use $\mathbf{E}$ as $\mathbf{X}$ and repeat the sequence given above. New test samples $\mathbf{X}_{test}$ can be corrected with the regression matrix $\mathbf{W}$ and loading matrix $\mathbf{P}$ of the calibration model by $\mathbf{X}^*_{test} = \mathbf{X}_{test} - \mathbf{X}_{test}\mathbf{W}(\mathbf{P}^T\mathbf{W})^{-1}\mathbf{P}^T$.

At each wavelength point within a given processing window, an individual OSC correction is conducted. This processing continues through the entire range of wavelengths until every portion of the spectrum is filtered. Full details of the piecewise OSC algorithm and its use can be found in *(12)*.

3.2 Performance of POSC and OSC on Representative NIR Data

We have compared OSC and POSC on several data sets that show problems of the sort described above (12). Here, we focus briefly on a selected result from the POSC filtering of the Cargill corn data set, one with unstabile baselines and with large amounts of scatter and other sources of extraneous variance. Figure 1 shows a comparison of POSC and OSC filtering prior to PLS modeling of protein for instrument mp6. orthogonality constraint by the local processing. Not only does POSC filtering lead to a significantly smaller PLS calibration model (a reduction in the PLS model of 4 latent variables for 1 POSC component removed), the resulting PLS model performs slightly better than a PLS model developed on unfiltered data in prediction of new samples. OSC filtering, on the other hand, merely decreases the PLS model by 1 latent variable from that produced on raw data and does not alter the predictive power of the PLS modeling. As there was 1 OSC component removed in filtering, there is no net benefit from OSC pre-processing, no matter which algorithm for OSC is used.

In Figure 1 and in all of the following discussion, the models are evaluated by considering the errors demonstrated in prediction of new data, either on separate (external) validation sets or in leave-one-out cross-validation. The error is calculated in terms of the root mean squared error of prediction or cross-validation (RMSEP or RMSECV) For example,

$$\text{RMSEP} = \left(\sum_{i=1}^{I} (y_i - \hat{y}_i)/I \right)^{1/2} \qquad h(1)$$

where y_i is the actual value for each run, $\hat{y}_i$ is the corresponding predicted value, and I is the number of samples in the external test set. The RMSECV is defined similarly; all that changes is the origin of the samples to be predicted. Generally, an improvement in RMSEP or RMSECV of about 0.1 units is indicative of real improvement in the modeling.

4. Localized Preprocessing with Wavelets for Model Improvement

Projection methods are ineffective when the interfering sources of variance in the calibration data are correlated to the property, either because of the design of the experiment or through accidental correlation, possibly through the action of noise. Wavelet analysis offers an alternative means to isolate and remove offending signals by their frequency composition – essentially their shape. This ability to deal with local aspects of the signal makes it feasible to remove interfering peaks whose shapes may be similar to those parts of the spectra believed relevant to the target while retaining the useful parts of the signal.

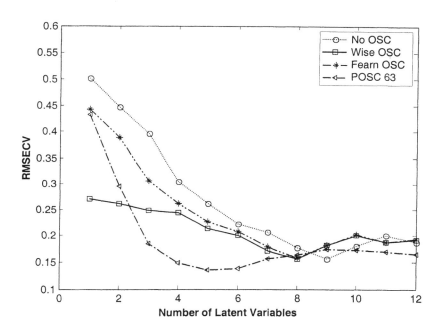

The results shown in Figure 1 confirm the benefit from releasing the Figure 1: RMSECV values of protein calibration in Cargill corn using PLS models with 1-12 LVs. 1 OSC or POSC component was removed. POSC used a 63 point window. Reproduced from reference 12. Copyright 2002 Elsevier Publishers.

Thus, wavelet analysis offers an alternative path for filtering "noise" from a calibration, strengthening the calibration and possibly improving its predictive powers (*13*).

According to wavelet theory, a discrete signal, f_i can be decomposed approximately as follows: $f_i \approx g_{i-1} + ... + g_{i-l} + f_{i-l}$, where f_{i-l} is the approximation component at the coarsest scale (level) l with a frequency no larger than 2^{i-l}. This component is orthogonal to the detail components g_j at different levels j (j = i-1, ···, i-l). These detail components are mutually orthogonal and their frequencies lie between 2^{j+1} and 2^j, respectively.

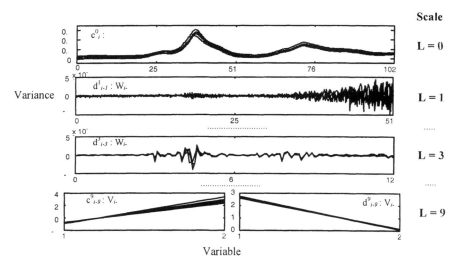

Figure 2: Discrete Wavelet Transformation of Gasoline Spectral Data. Note the compression of the spectra with scale

While many different approaches exist that use wavelets to process spectral data, we have found that the ones commonly reported in the chemometrics literature offer little real benefit in calibrations, mainly because of the compression associated with the discrete wavelet transform as conventionally implemented in toolboxes and elsewhere. Figure 2 shows the way in which a spectrum is decomposed with the conventional algorithm for the discrete wavelet transform (DWT). Note that the lowest frequencies in the spectrum (highest wavelet scales) are represented by a single point. While such compression is advantageous in image transmission and analysis, it is not helpful to signal processing data to be used in multivariate calibration because the low frequency effects, often the reason why calibrations fail, are inaccessible to much signal processing.

We have found that a version of the Mallat algorithm (*14,15*), where spectral signals are decomposed to a set of scales of constant length, is far more useful. The Mallat algorithm can be made to function much like a prism, in that it partitions a complex spectrum according to the component frequencies, so we call the new algorithm the "wavelet prism" (WP) (*16*). Baseline-like information can be found in the lowest-frequency approximation components, e.g., f_{10}, whereas noise-like information is mostly located in high-frequency detail components, e.g., g_1 and g_2.

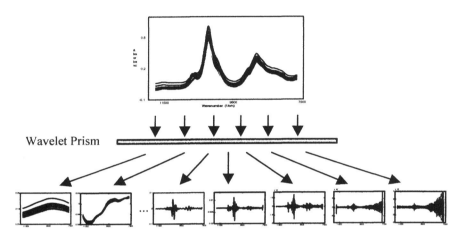

Figure 3: Mallat "Wavelet Prism" Transform of Data from Figure 2. Scale decreases from left to right, but all spectra are the same length. Reproduced from reference 16. Copyright 2002 John Wiley and Sons, Ltd.

Therefore, it is possible to strip out baseline (for example) by removing the lowest-frequency approximation component, if the background, analytical signal and noise contributions do not interact with each other and if the spectrum f satisfies the linear additivity rule $f = g_n + g_s + f_b$, where g_n is the detail component assigned to noise contributions, g_s is the detail components assigned to signal and f_b is the approximation components assigned to noise. If the scales representing background, signal and noise can be identified, those scales describing signal can be kept, those describing undesirable effects can be removed, and finally the wavelet reconstruction can then be run, leading to a shape-filtered signal. If the background and noise components can be isolated in this fashion, the filtering can lead to a smaller, cleaner calibration model, improving the life and utility of the calibration.

As an example of the effects of wavelet removal of background, we consider analysis of a NIR dataset (*16*). We rearranged the calibration and prediction sets of that dataset to create a system with a controlled, varying background response (produced by the water component) and a known analyte (acetone). In our analysis, we focus only on the acetone band near 2100 nm, where the large signal from water overlaps the weaker acetone band because this system simulates a small analyte signal on a large, varying background, yet is easily controlled and reproduced. The rearranged calibration and prediction sets allow us to explore the cases where the background in new data is not commensurate with that seen in the calibration step. These cases should lead to failure of a conventional calibration models but not models made robust to external effects, which is the result seen in Table I.

Table I. Calibration and prediction errors for PLS calibration of acetone from NIR spectra with varying, uncalibrated background

Baseline correction methods	*Latent variables*	*Cross-validation (%, w/w)*		*Prediction (%, w/w)*	
		RMSECV	*Bias*	*RMSEP*	*Bias*
None	6	1.173	-0.037	4.310	3.289
1st Derivative	5	1.033	-0.032	4.452	-3.515
Multiplicative Scatter Correction	3	0.829	-0.021	7.622	-6.049
Orthogonal Signal Correction	5	0.612	-0.109	15.284	-8.321
Polynomial fit method*	2	1.361	-0.009	13.491	-10.93
Wavelet baseline removal	2	3.171	-0.238	3.207	-0.956

*The baseline removal method available in Eigenvector Technology's *PLS Toolbox*. Reproduced from reference *16*. Copyright 2002 John Wiley and Sons, Ltd.

Interestingly, several popular background correction methods appear to give satisfactory results in cross-validation, but fail spectacularly in the prediction of the new data, a consequence of the incomplete and inconsistent removal of background by these methods and the failure of the PLS models built on their output to generalize well. The wavelet-directed removal of background, while seemingly less successful in cross-validation, shows no difference in performance on new data, and it permits construction of a much smaller calibration model, with only 2 latent variables. Most of the information (variation) for the raw NIR spectra of acetone is contained in the range of levels

3 to 8, which implies a multiscale nature for the NIR spectra. Wavelet preprocessing is successful here because the broad water background is mostly confined to scales 10-12, making possible quantitative separation of acetone signal from the water background by the wavelet preprocessing step.

5. Robust Modeling through Multi-Scale Calibration

While preprocessing helps to create more robust models by upweighting of property-related signals and downweighting of interferent effects, it is possible to lose some useful information during the preprocessing by projection, wavelet filtering or de-noising, since the properties' contribution to the spectrum may be small but non-zero for large sections of the frequency domain and the noise that correlates interferences to the property may also decrease some correlated signals as well. This information would be lost in a removal of components that might seem to be contributing to baseline or noise but might also be of vital importance for improving predictive regression models. Similarly, when the analyte signal is multiscale in nature, selection of single wavelengths, or of bands of wavelength, also loses information though the truncation process *(17,18)*. An alternative to component removal is needed to improve model robustness without loss of information on the property of interest.

It is known that the statistical characteristics (e.g., the moments) of the raw spectra remain the same at each frequency scale after wavelet decomposition, because a linear transform such as the wavelet transform conserves them. Therefore, the frequency components of spectra obtained by a wavelet decomposition using the WP algorithm described above may be modeled separately at different frequency scales, if the linear correlative relationship between the raw spectra and the target property can be statistically described. As a result, it should be possible to implement regression analysis on *dual-domain* (i.e., wavelength-frequency) spectra over the entire wavelength and frequency domains at the same time, and unlike wavelength selection methods or the wholesale removal of scales to strip background from signals, this modeling should not cause significant information loss because no truncation is involved.

The regression analysis on a dual-domain spectral set is a two-step procedure, done in a way similar to that used for regular (single-domain) regression methods *(17,18)*. The first step is to establish a dual-domain regression model in a calibration of the $m \times 1$ dependent vector **y** (the property) on a set of independent variables contained in a multi-scale spectral tensor $\underline{\mathbf{X}}$ {$\mathbf{X}_k$, k = 1, 2, ..., *l*+1}. The second step is to predict values for the new properties based on a prediction set $\underline{\mathbf{X}}_u = \{ \mathbf{X}^T_{1,u} .... \mathbf{X}^T_{l+1,u} \}^T$.

Consider the combination of a set of single-domain regression models

$$y = \sum_{k=1}^{l+1} \mathbf{X}_k \boldsymbol{\beta}_k + e \qquad E(e) = 0, \qquad \text{Cov}(e) = \sigma^2 I \qquad (2)$$

where $\boldsymbol{\beta}_k$ is the $p \times 1$ regression coefficient vector for the frequency component at the kth scale in the single-domain spectra, $\mathbf{e}$ denotes an $m \times 1$ error vector, and $E(\cdot)$ and $Cov(\cdot)$ are the expectation and covariance, respectively. The goal of the multi-scale regression analysis is to calculate the matrix of regression coefficients $\mathbf{B} = \{\boldsymbol{\beta}_1, ..., \boldsymbol{\beta}_{l+1}\}$ with the lowest associated error of prediction when applied to new samples spanned by the calibration. While any regression method could be used in building the regression relationship, we have considered PLS and PCR to date (*19*).

Exact solution of the dual-domain calibration defined in equation (2) for the optimal regression model defined there is not straightforward, but satisfactory performance may be obtained by an approximate solution involving a weighted average of the *l*+1 single-scale calibration models. For each of the *l*+1 regressions, the scalar g_k defines the weight for the *k*th single-domain calibration model, as determined by cross-validation of the calibration set according to $g_k = s_k^2 / \sum_{k=1}^{l+1} s_k^2$, where s_k is the reciprocal of the cross-validation error in the calibration set. To see how an approximate solution may be obtained, consider dual-domain regression using PLS. Analysis using PCR is similar. In the case of PLS, a separate PLS regression on each frequency component (scale) of the dual-domain spectral tensor $\underline{\mathbf{X}}$ is first performed with respect to the dependent vector $\mathbf{y}$, and the *l*+1 PLS regression vectors obtained are then weighted according to their relative predictive ability for the target, as judged by cross-validation. The frequency component with the highest predictive relationship to the analytic target will gain the highest weight, and the one with the least predictive relationship (where information is minimal) will receive a correspondingly small, but non-zero, weight. In this way, all scales contribute to the overall regression model, and all wavelengths also contribute, hence the term dual-domain regression.

In the prediction step with the PLS dual domain model, an unknown sample $\mathbf{x}^T_u$ is first decomposed by the WP algorithm, followed by multiplication of the frequency components $\mathbf{x}^T_{k,u}$ ($k = 1,2,...l, l+1$) with the weighted kth regression vector according to the relation $\hat{y}_u = \sum_{k=1}^{l+1} x^T_{k,u} \hat{\beta}_k$.

Because the weighting of the regression defined in equation (2) combines the sets of latent variables generated from the separate analyses of the wavelet decompositions at different scales, this defines a multiscale regression model combining information from both the wavelength and frequency (scale) domains, but there will be only a single set of latent variables produced from dual-domain PLS or PCR, just as in conventional PLS or PCR. However, the weighted latent variables produced by dual-domain PCR and PLS, in general, will differ from those produced by conventional PCR and PLS, respectively, because of the weighting of the sets of latent variables for the different scales.

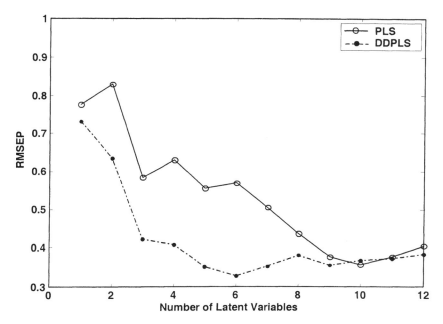

Figure 4: Starch prediction of Cargill corn samples measured on instrument mp5 using several PLS models. Modified from reference 19. Copyright 2003, John Wiley and Sons, Ltd.

The multiscale analysis offers real advantages over conventional calibration, as Figure 4 shows. There, dual-domain PLS regression (DDPLS) is compared with conventional PLS regression (*19*). The DDPLS shows better performance because of the multi-scale advantage gained from the parallel analyses. Interestingly, PLS regression done on all wavelet coefficients at once (*19*) or PLS regression done on DWT coefficients (*20*) loses the multi-scale benefit.

Dual-domain regression offers many of the advantages of local methods while retaining global information. We are exploring ways of improving the multiscale nature of the method and of applying the method to strengthen multivariate calibrations.

Acknowledgements

This work was supported by the Center for Process Analytical Chemistry.

References

1. Chu, X.L.; Yuan, H.F.; Lu, W.Z. *Spectrosc. and Spec. Anal.* **2001**, *21*, 881-885.
2. Fearn, T. *J. NIR Spectrometry* **2001**, *9*, 229-244.
3. Feudale, R.N.; Woody, N. A.; Myles, A. J. ;Tan, H.-W.; Brown, S.D. ; Ferré, J. *Chemom. Intell. Lab. Syst.* **2002**, *64*, 181-92.
4. Adhihetty, I.S.; McGuire,J.A.; Wangmaneerat, B.; Niemczyk, T.M. Haaland,D.M. *Anal. Chem.* **1991**, *63*, 2329-2338.
5. Swierenga, H.; Haanstra, W.G; de Weijer, A.P.; Buydens, L.M.C. *Appl. Spectrosc.*, **1998**, *52*, 7-16.
6. Swierenga, H.; de Weijer, A.P.; van Wijk, R.J.; Buydens, L.M.C. *Chemom. Intell. Lab. Syst.* **1999**, *49*, 1-17.
7. Committee E13-11, American Society for the Testing of Materials, Method E1655-00, *ASTM Standards* **2002**, *03.06*, 573-600.
8. Wold, S.; Antti, H; Lindgren, F.; Öhman, J. *Chemom. Intell. Lab. Syst.* **1998**, *44*, 175-185.
9. Fearn, T. *Chemom. Intell. Lab. Syst.* **2000**, *50*, 47-54.
10. Fernández Pierna, J.A; Massart, D.L.; de Noord, O.E.; Ricoux, Ph. *Chemom. Intell. Lab. Syst.* **2001**, *55*, 101-108.
11. Westerhuis, J.A.; de Jong, S.; Smilde, A.K. *Chemom. Intell. Lab. Syst.* **2001**, *56*, 13-25.
12. Feudale, R. N.; Tan, H.-W.; Brown, S. D. *Chemom. Intell. Lab. Syst.* **2002**, *63*, 129-38.
13. Bakshi, B.R. *J. Chemom.* **1999**, *13*, 415-434.
14. Mallat, S.G. *Trans. Am. Math. Soc.* **1989**, *315*, 69-87.
15. Mallat, S.G. *IEEE Trans. Pattern Anal. Machine Intell.* **1989**, *11(7)*, 674-693.
16. Tan, H.-W. ; Brown, S.D. *J. Chemom.* **2002**, *16*, 228-40.
17. Tan, H.-W.; Mittermayr, C.R. ; Brown, S.D. *Applied Spectrosc.* **2001**, *55*, 827-33.
18. Alsberg, B.K.; Woodward, A.M.; Winson, M.K.; Rowland, J.J.; Kell, D.B. *Anal. Chim. Acta* **1998**, *368*, 29-44.
19. Tan, H.-W.; Brown, S.D. *J. Chemom.*, **2003**, *17*, 111-122.
20. Trygg, J. ; Wold, S. *Chemom. Intell. Lab. Syst.* **1998**, *42*, 209-220.

Chapter 3

Interpretation and Validation of PLS Models for Microarray Data

Fredrik Pettersson and Anders Berglund

Research Group for Chemometrics, Department of Chemistry,
Umeå University, Umeå, Sweden

Introduction

By the use of microarray technologies gene expression can be monitored on genomic scale in a high throughput manner. The full power of these techniques has been unleashed, as the sequence information from large genomic sequencing projects such as HUGO has been deposited. An important advantage of microarray experiments is that the relative low cost and ease of use makes it practical to do detailed and systematic studies (1-3). The number of potential genes in a genome easily reaches many thousands. With microarray analysis the global expression of these genes can be monitored on a single chip in one single experiment.

Each microarray study results in a data matrix where the number of variables (genes) far exceeds the number of observations (experiments). Microarray expression data is often ill conditioned as it is noisy and variables are collinear with a high degree of missing values. Poor reproducibility is a main issue. The problem is no longer to obtain the data but to extract the knowledge that is embedded in the data. As classical statistical methods such as pair wise t-tests are not well suited for this kind of data structures a demand for new better-suited techniques is emerging. Today techniques such as hierarchical clustering, k-means clustering and self-organizing maps are used for analyzing microarray expression data.

The use of microarrays has in many ways revolutionized gene expression analysis with a large number of interesting applications. The data mining strategy used to interpret microarray data depends on the experimental design and can be broadly divided into two categories: coordinated gene expression and differential gene expression. Coordinated gene expression analysis involves the assessment of a large number of genes over a period of time such as variation during the cell cycle. The differential gene expression approach generally consists of pair wise comparisons between normal/abnormal samples. We will focus on the latter in this paper.

Microarray expression analysis is a powerful tool in functional genomics where the aim is to assess functional properties of genes. The basic assumption that genes with similar patterns of regulation over different conditions have similar function and may be involved in the same pathways is generally used to assign functional properties to unknown genes. This information can then be used to construct regulatory networks, to find disease-associated genes (potential drug targets) and to find interacting components.

Another important application is to perform expression profile based disease classification/diagnosis. Diseases with similar symptoms may have different underlying mechanisms and should be treated according to the underlying mechanism and not the symptom. Tumors with similar histopathological appearance can follow significantly different clinical courses and show different responses to treatment. Disease classification by analyzing expression profiles shows a new way of determining the correct treatment and thereby increasing the success rate and avoiding toxic side effects. A lot of efforts have been put especially in the case of cancer class prediction. The expression data is highly complex with many independent sources of variation. The success of this approach is highly dependent on the development of clustering- and classification-methods that can deal with this data with highest possible specificity and sensitivity. Except for being accurate it is also important that the analysis methods are fast and interpretable both statistically and biologically. The methods applied today perform reasonably well but not without some misclassifications.

PLS, partial least squares, has shown to be a powerful method for performing multivariate statistical analysis on similarly conditioned datasets in the field of chemometrics (4). PLS has recently been applied for studies of coordinated (5) as well as differential geneexpression (6). As previously mentioned microarray data often is very noisy due to the experimental procedure and consists of a large number of variables that far exceeds the number of observations. When analyzing this kind of data there are two obvious risks. It is easy to obtain overfitted models with poor predictability. With a huge number of variables there is an increasing risk of getting false positives just by chance. In this work we investigate how to check if a PLS model is overfitted or not and how to determine which genes that are most important for discriminating between biological samples of different types. One important step is to define a cut-off value where one can say that all the genes above the value is correlated to the response with a certain degree of confidence.

Experimental

In this project we have analyzed a well-studied dataset including expression values for 7070 genes for 72 leukemia samples. The experiments were performed by Golub *et al.* (7). The samples were prepared and categorized by histopathological appearance and marker specific recognition at collaborating hospitals. The samples were taken from bone marrow or peripheral blood and classified as type ALL (lymphoid origin) or AML (myeloid origin). The data were divided into a modeling set consisting of 38 samples and an independent test set consisting of 34 samples. The chips used were of Affymetrix type and the gene expression levels pre-normalized. Nguyen *et al.* have earlier also analyzed this data set with PLS. (6)

Methods

PLS-DA

In PLS-DA dummy variables are used for describing the class belonging of the different samples. This is done by creating binary variables, one variable for each class, with for example ones and zeros, where a one represents that the object belongs to that class. With a PLS-DA model it is possible to predict class belonging by looking at the predicted class variable. A value above 0.5 means that the specific sample belongs to that class and below 0.5 that it does not belong to that class.

Generation of a ranking list

When we know that we have a good model and we can make predictions the next interesting step is to investigate which genes differentiates between ALL and AML samples. The obvious way is to look at the regression coefficients from the PLS model and rank them according to their size. If a gene has a large regression coefficient with AML as a response it means that the gene is upregulated for AML and vice verse for ALL. The problem with the regression coefficients is that they are affected by the variation in the X-matrix that is not correlated to the response (8). The O2-PLS algorithm by Trygg *et al.* gives a PLS model which is only describing the variance that is correlated to the response. This is done by first removing the non-corrected variance in the X-matrix. Another way to remove the variation in the X-matrix is to use some of the OSC-algorithms that are available, but they are not as simple to use as the O2-PLS algorithm.

Trygg *et al.* (8) also showed that, for a single PLS model, the first weight vector of a PLS model, **w1**, is the best estimate of how important a variable is for describing the response. Later components are only needed for correcting the predictions made by the first component for all the variation in the X-matrix that is not correlated to the response but still affects the prediction. The **w1** vector has with success been used for finding genes with cell cycle-coupled transcription (5). Another alternative for ranking the gene importance would be to use the VIP value which is a summation of the absolute value for all the w-vectors in a PLS model taking into account how much each components explains. This value suffers from the same problem as using the regression coefficients, it takes into account all the components in the PLS model.

Finding a cutoff value

The next step, after making a ranking list of the importance each gene, is to find a cut-off value where one can rank, with a certain degree of confidence, all the genes above the cutoff as correlated to the response with a certain degree of confidence. This can be done by, for example, using random genes for finding a proper cut-off value. If the genes are only permuted, the same error distribution and variance is kept whereas the correlation to the response is changed. This approach was used for finding a cut-off value for the number of genes with cell cycle-coupled transcription (5).

A PLS model is calculated using the randomized data set and the w1-vector is used for finding an appropriate cut-off. To avoid skews when comparing the results the weight vector is not normalized to unit length as it generally is, since normalization would increase the weights for the simulated genes.

The importance of the significance level of the regression can be illustrated by considering the last candidate gene in the ranking list. If the expression profile of this gene is not dependent on ALL or AML the probability of scoring at least this high, purely by random chance, is $(100*\alpha)$ percent. That is, the p-value for the specific gene is α. Conversely, the probability that the gene's expression is genuinely related to ALL or AML is approximately equal to (1- α). The smaller the value of α, the fewer false positives there will be among the candidate genes, but also the more false negatives there will be among the genes below the threshold.

Results and Discussion

PLS-DA model for ALL/AML

Figure 1 shows the two first t-scores plotted against each other for a PLS-DA model describing the differences between ALL and AML. A separation between the two types is seen in figure 1, both for the modeling set and the test set. That the separation is seen also in the test set is assuring, since these samples have not been part of the model and thus the separation is not a consequence of an overfitted model. This is always a problem when there are many thousand of variables and relatively few observations. The predictions of the response for the test set samples shows that all test set samples, as well as the modeling set samples, are predicted correctly.

An external test set is the best way to evaluate the predictive power of a model but it is not always possible to have an external test, below we will show that there are alternative ways to evaluate the predictive power of a model.

The PLS model with three components explains 95.8% (R2Y) of the variance in Y, with a corresponding cross-validated (7 cross-validation groups) value of 80.6% (Q2). Only 33.4% of the variance in the X-matrix are used (gene expressions) which indicates that most of the genes are not related to the response (AML-ALL separation) at all.

Validation of Q2 and R2Y

These values can be further validated by a permutation test where the response is permuted and new models are calculated using the permuted response . If R2Y and Q2 are equally high as the original model the statistical values for the model is not trustworthy since even a random response and, hence

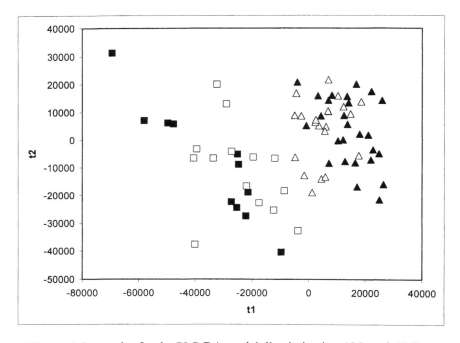

Figure 1 Score plot for the PLS-DA model discriminating ALL and AML samples. ● – ALL modeling set, ○ – ALL test set, ■ – AML modeling set, □ – AML test set

a random model can give equally high values. Figure 2 shows the results from such a permutation study where 50 permutations have been made. One can see that the R2Y value is equally high for all the permuted models as it is for the original model, top right. This is not surprising since there are so many variables, and few objects, there will always be some variables that are correlated with the response. The Q2 at the contrary, shows a low value for all the permuted models. This indicates that the high Q2 for our model is not due to a chance correlation.

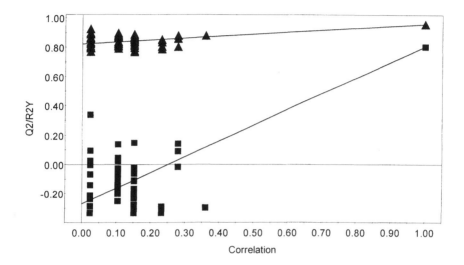

Figure 2 Validation of the Q2 and R2Y values from the PLS model for ALL and AML. The x-axis shows the correlation between the original y-response and the permuted y-response. The box and triangle up to the right are from the original non-permuted model while all the others are from the models with a permuted y-response. Notable is that all models have a high R2Y but only the original model has a Q2 higher than 0.8. ■ – Q2 ▲ – R2Y

Comparing w1 and regression coefficients

Figure 3 shows the coefficients from the PLS model plotted vs. the first weight vector w1. As figure 3 shows there is a correlation between these two, but also that there are genes that a deviating from the line. Thus, if a ranking list would be generated from these two different vectors, the ordering of the genes would not be the same.

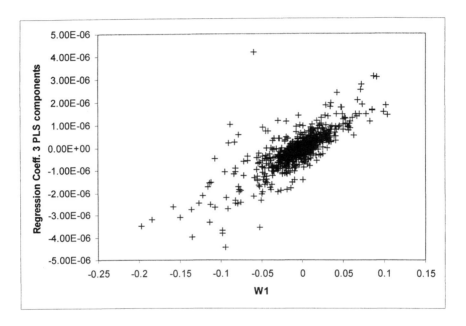

Figure 3 The regression coefficients from the PLS model with 3 components vs. the first weight vector, w1.

Fnding a cut-off value

A new X-matrix was created by column-wise permutation of the elements. This was repeated ten times so that the resulting randomized X-matrix consists of 70700 variables. This was both done for the modeling set and for the whole data set where the modeling and the test set was merged. By merging the data sets we get more observations in the resulting dataset and it is possible to get more genes above the threshold value.

Using our cut-off strategy, with a significance level of 0.5%, 122 genes scored above the threshold, for the modeling set. If all the observations are used, both from the test and the modeling set, 177 genes scores above the threshold.

Comparing the top 122 genes from the modeling set and the top 177 genes from the merged dataset reveals that 100 of the 122 genes are also present in the 177 list. Thus, there is not a perfect overlap between the two lists since 22 genes from the 122 list are not present in the 177 list. This imply that when we add more observations, the ranking list also changes, some genes get more importance and vice versa. Further, the first 34 genes in the 122 ranking list are among the 177 genes from the ranking list based on all observations.

This difference can either be related to the selection of to many genes, genes with no differencing power between ALL and AML samples, or that the addition of more samples actually adds new information which is not present in the modeling set.

Discussion and Conclusion

In this work we have successfully used a multivariate projection method, PLS-DA, to perform molecular classification of leukemia samples. We have obtained predictive models with the ability to discriminate between tumors of myeloid or lymphoid origin with no misclassifications. PLS-DA is highly useful for obtaining predictive models with the ability to discriminate between different types of samples still we have to be cautious so that models are not overfitted and that genes do not end up high in a ranking list just by coincidence. Traditionally when analyzing PLS models variables are ranked by their regression coefficients. This may be misleading since systematic variation in the X-matrix unrelated to the response is included in the regression coefficients. Another option is to rank the genes by their **w1** values. Our study shows that genes are differently ranked using the two different ways of interpreting the variables. We have also presented a way of defining a cut-off value where one can say that all of the genes above the cutoff value in a ranking list is correlated to the response with a certain degree of confidence. When using all observations from both the test and modeling set 177 genes scored above the threshold value at a significance level of 0.5%. Models can be validated by permuting the dataset so that response values are randomly shifted between the different observations. All permuted models show poor predictive properties (Q2) while the correlation (R2Y) is always high. This is a consequence of the large number of genes (7070).

References

1. Fodor, S. P.; Rava, R. P.; Huang, X. C.; Pease, A. C.; Holmes, C. P.; Adams, C. L. *Nature*, **1993**, *364*, 555-6.
2. Lander, E. S. *Nature Genetics*, **1999**, *21*, 3-4.
3. Schena, M.; Shalon, D.; Davis, R. W.; Brown, P. O. *Science*, **1995**, *270*, 467-70.
4. Wold, S.; Ruhe, A.; Wold, H.; Dunn III, W. J. *SIAM J. Sci. Stat. Comput.*, **1984**, *5*, 735-743.
5. Johansson, D.; Lindgren, P.; Berglund, A. *Accepted in Bioinformatics, 2003, vol 19*, **2002**.
6. Nguyen, D. V.; Rocke, D. M. *Bioinformatics*, **2002**, *18*, 39-50.
7. Golub, T. R.; Slonim, D. K.; Tamayo, P.; Huard, C.; Gaasenbeek, M.; Mesirov, J. P.; Coller, H.; Loh, M. L.; Downing, J. R.; Caligiuri, M. A.; Bloomfield, C. D.; Lander, E. S. *Science*, **1999**, *286*, 531-537.
8. Trygg, J. *Journal of Chemometrics*, **2002**, *16*, 283-293.

Chapter 4

Chemoinformatics: Perspectives and Challenges

Ling Xue[1], Florence L. Stahura[1], and Jürgen Bajorath[1,2,*]

[1]Department of Computer-Aided Drug Discovery, Albany Molecular Research, Inc. (AMRI), AMRI Bothell Research Center (AMRI-BRC), 18804 North Creek Parkway, Bothell, WA 98011
[2]Department of Biological Structure, University of Washington, Seattle, WA 98195

> This chapter discusses some aspects of chemoinformatics as an evolving discipline in the context of life science research. Highlighted topics include selected scientific questions and also more strategic issues such as the unification of diverse informatics approaches or the formation of viable interfaces with experimental programs. It is hoped that this chapter might aid in putting some of the current trends in the chemoinformatics field into perspective.

The term chemoinformatics was probably first introduced in the literature in 1998 by Frank Brown (*1*). Although this might indicate that we are looking at a very young discipline, many -but not all- of the activities that are currently categorized under the chemoinformatics label have a longstanding history in computational chemistry research. The dominant theme of informatics-driven activities in life science research is the transformation of rapidly growing amounts of biological and chemical data into knowledge. However, this general theme covers (in chemistry, biology, and pharmacology) a diverse array of research and development activities and different applications.

© 2005 American Chemical Society

What Is It?

Similar to the situation in biology, the advent of high-throughput technologies in chemistry in the mid 1990s has been a driving force for the introduction and increasing popularity of chemical informatics. Whereas in biology the analysis of large amounts of DNA sequence data became a critical issue, combinatorial chemistry and the ensuing needs for systematic synthesis planning and compound library design and management triggered the development of informatics tools. However, chemoinformatics, as we define and understand it today has at least two major roots, high-throughput chemistry technologies and, in addition, the qualitative and quantitative computational analysis of structure-activity relationships of small molecules (2). Accordingly, the present spectrum of chemoinformatics approaches includes compound registration and database management, reaction and library design, molecular similarity and diversity analysis, and all methods that correlate structural features, physico-chemical properties, and biological activities of compounds, regardless of their sources (1,2). Currently available structure/property-activity methodologies are highly diverse and include different types of clustering and database mining methods, multi-dimensional QSAR (quantitative structure-activity-relationship) approaches, similarity search tools, and virtual (or *in silico*) screening approaches (3). In addition, statistical methods designed to analyze HTS (high-throughput screening) data and derive predictive models of biological activity are also regarded as a part of chemoinformatics (3). Moreover, the prediction of (*in vivo*) ADMET (absorption, distribution, metabolism, excretion, toxicology) characteristics of active compounds is often considered a chemoinformatics approach, although methods employed in this context range from statistical analysis and QSAR-type approaches to quantum-chemical calculations and modeling of enzyme reactions (4).

There is no doubt that it is often difficult to say where chemoinformatics research begins and ends. The boundaries to other computational disciplines are ambiguous, which is not surprising for an evolving R&D (research and development) area. However, as discussed later on, rigidly distinguishing between different informatics-driven research fields may indeed not be very meaningful. It is also worth noting that many of the current activities in biological and, in particular, chemical informatics have evolved in drug discovery settings. Thus, in order to appreciate developments and trends in this field, it is often helpful to also reflect on the current situation in pharmaceutical research.

Examples of Popular Topics

What are some of the current focal points of chemoinformatics R&D? In the pharmaceutical industry, the development and continuous improvement of efficient, flexible, and relational infrastructures for compound registration, acquisition, and management, including Web-based systems, continues to be an important task, just as it has been from the beginning. Essentially any pharmaceutical or chemical company and also any chemistry department in academia are faced with the challenge of developing, analyzing, and maintaining compound databases of dimensions never seen before, often containing literally millions of structures. Thus, there is little doubt that the need for efficient and easily accessible and exchangeable database structures and management systems will further grow.

On the research front, much effort in the field has recently been dedicated to the systematic analysis and comparison of molecular characteristics and property distributions in different databases. Many of these studies aim to improve our understanding of what renders a molecule drug-like or try to systematically distinguish between drugs and non-drugs (5). Furthermore, a variety of different database mining and virtual screening methods have been introduced in recent years to effectively search for active compounds (3). Similar to the situation in database management, a major challenge for these methods has been the dramatically increasing size of compound sources. This requirement alone significantly influences method development activities. For example, very large numbers of database compounds ("real" and/or virtual) prohibit the application of classification methods that rely on pair-wise compound or distance comparisons in chemical space (such as conventional clustering techniques) and have spurred on the development of currently popular low-dimensional partitioning methods (3,6).

Current Trends

A few general trends can also be observed that have begun to influence the direction of R&D efforts in the chemoinformatics arena. It has been becoming increasingly clear over the past few years that high-throughput technologies alone cannot be expected to revolutionize drug discovery and significantly increase its ultimate output (7). Accordingly, chemistry has begun to depart again from a pure numbers game ("make more compounds – screen them faster") and focus more on knowledge-based approaches. For example, whereas the design and generation of diverse combinatorial and large virtual libraries dominated many computational efforts in the field early on, the emphasis has now shifted to designing smaller but "smarter" (e.g., enriched with drug-like molecules) and target-focused compound libraries (2).

It has also become clear that achieving desired biological activity and potency of hits or leads is necessary but not sufficient to produce high-quality drug candidates. This is illustrated by the dramatic attrition rates of compounds during clinical trials, which is currently probably the major bottleneck in drug discovery (2). This situation has influenced chemoinformatics R&D and is responsible for the significantly increasing interest in the development of computational ADMET concepts (4,8).

Challenges

As one might expect, taking into account the rapid growth chemoinformatics has experienced over just a few years, the discipline also faces a number of problems and critical issues that could either substantially hinder further progress or, if successfully resolved or circumvented, provide a wealth of new opportunities (8). Some of these difficulties can be attributed to the fact that much of the chemoinformatics work is currently done in drug discovery environments. An obvious (albeit understandable) "strategic" drawback of this situation is that many of the developments and findings are kept proprietary, at least for an extended period of time. These include, among others, internal database structures and information systems and, equally -if not more- important, the results of computational screening or lead identification programs and the experimental evaluation of focused libraries. A consequence of this situation is that the literature in the area of chemical information and informatics is currently dominated by publication of novel methods, whereas reports of practical (drug discovery) applications are the exception, rather than the rule. However, it is worth noting that specific journals have become a forum for chemoinformatics R&D, in particular, the *Journal of Chemical Information and Computer Sciences* for method development and the *Journal of Medicinal Chemistry* for drug discovery applications and case studies. This makes it in general difficult to judge about the true performance of chemoinformatics approaches and put them into scientific perspective. In order to establish generally acceptable standards for the comparison of different methods, the chemoinformatics field would indeed greatly benefit from the availability of more "real life" examples and the disclosure of more high-quality datasets (e.g., HTS data or specifically designed compound libraries). In addition to this rather general aspect, there are also a number of specific challenges.

Unresolved Scientific Questions

Despite the fast-paced development of different informatics tools, some fundamental scientific issues remain to be understood in order to increase the accuracy of computational methods and their impact on chemistry programs (8). Among these, the difficulty to reliably predict *in vivo* properties of clinical candidates is probably one of the scientifically most complex issues. Despite significant interest in these approaches, computational progress is hindered by the fact that many ADMET-relevant physiological effects are themselves not yet well understood. Thus, since multiple effects usually influence the *in vivo* fate of compounds, it is in general difficult to assemble accurate and sufficiently large compound training sets for various ADMET predictions. Consequently, much remains to be learned until sound and more widely applicable computational models can be built.

Other critical questions are much more chemistry-specific. For example, for compound design, the current inability to accurately predict binding energies and affinities, regardless of whether they are calculated from QSAR-type analysis or simulations of protein-ligand complexes, continues to be a major caveat. Furthermore, limited synthetic feasibility of many computer-designed molecules continues to present a major difficulty for the integration of informatics and chemistry efforts, despite the fact that various programs for de novo compound design are already available of for at least a decade. Thus, the development of predictive methods to increase synthetic success rates of designed molecules would also be an important step forward. To give another example, uncertainties in calculating bioactive three-dimensional molecular conformations have a significant negative impact on 3D database and pharmacophore search methods and 3D QSAR techniques, the latter being a cornerstone methodology for lead optimization programs. In all of these instances, there is significant room for improving the scientific basis and accuracy of chemoinformatics approaches.

It might also be mentioned that the boundaries between molecular modeling, which is long established as a computational discipline, and informatics have become rather fluent. For example, knowledge-based modeling approaches make use of rapidly growing databases such as the Protein Data Bank or the Cambridge Crystallographic Database, which often requires the application of advanced data mining techniques. In addition to biological and chemical informatics, molecular modeling significantly adds to the spectrum of computational approaches that are available for the life sciences.

Education

A different yet equally critical issue for the future of the young chemoinformatics discipline is the fact that curricula and teaching programs that merge computer science and chemistry still need to be established. However, current developments are promising. The first graduate and postgraduate courses in chemical informatics have been introduced in the US and Europe (9), or are at least in the planning stage. Similar to the situation in bioinformatics (at least until very recently), the majority of scientists who currently work on chemoinformatics problems come from different backgrounds and many have more or less trained themselves. It is anticipated that the demand for young scientists who can successfully operate at the interface between computer science and chemistry will substantially grow in the coming years. Thus, academia is challenged with providing relevant educational opportunities in this area, either as part of conventional chemistry studies or standalone programs.

Opportunities

Any of the general or more specific challenges that chemoinformatics faces today (as discussed above) would, if successfully tackled, bring the field a significant step forward and further improve the impact on life science research. In addition, some other areas can be pointed out that might provide some significant opportunities for growth. Without doubt, more progress will be made in the development of algorithms and methodologies for chemoinformatics applications, considering the substantial resources that are being dedicated to these efforts. These developments will be supported by the availability of faster computers, efficient clusters, and essentially unlimited storage space. As databases structures and chemical information systems mature, deriving knowledge from raw chemical data will also become easier. However, beyond algorithms and databases, other significant future opportunities exist in closely integrating chemoinformatics and experimental programs (*2*) and in merging different informatics disciplines in the context of life science research.

Interfaces with Experimental Research

If closely interfaced, chemoinformatics tools and approaches can have a substantial impact on established experimental disciplines, at least in drug discovery. Although a number of pilot studies and programs already exist that highlight the success of such integration efforts (*2*), there is much room for improvement. In order to illustrate some of these opportunities, we will discuss two examples.

Virtual and High-Throughput Screening

Despite the many technical advances in HTS towards even higher screening throughput, it is increasingly being recognized that virtual screening (VS) and other chemoinformatics tools can significantly aid in improving the hit rate of HTS programs, reducing the number of compounds to be screened (either by building predictive models or by focusing of screening libraries), and in rationalizing the results (*10,11*). This is perhaps best exemplified by the focused or sequential screening paradigm where HTS and VS are used in an iterative manner (*10,11*). In this case, smaller subsets are selected from source libraries by *in silico* screening and subjected to HTS. The results are then analyzed and used to refine the VS approach for the next iteration, and the process is continued until a sufficient number of high-quality hits are obtained. Figure 1 illustrates this process and shows different informatics components that aid in sequential screening.

It is likely that the need for informatics-driven iterative screening protocols may soon increase, if the numbers of potential therapeutic targets and the sizes of compound libraries continue to grow at present rates. However, for chemoinformatics, the interface between HTS and VS provides challenges and opportunities beyond VS methodologies. For example, for iterative screening to be practical, highly flexible compound registration, handling, and database management systems

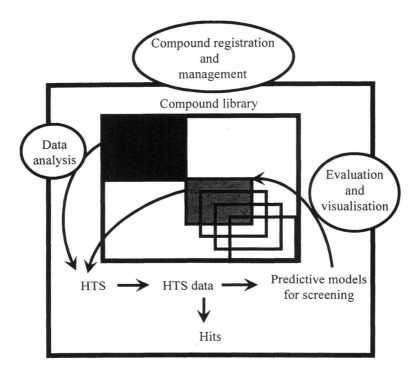

Figure 1. Iterative screening by combining HTS and VS. For VS, many different methods are available including compound clustering and partitioning or similarity searching (11). Following the iterative screening approach, small library subsets (in grey) are computationally selected and screened until a sufficient number of hits are obtained. During these iterations, the VS calculations are continuously refined. Predictive models of biological activity can also be derived from partial HTS dataset (in black) and used as a basis for subset selection. Other chemoinformatics components required to support the sequential screening process are circled.

must be implemented so that individual compounds can be cherry-picked and assembled from many different screening plates. In addition, requirements for efficient analysis and visualization of assay data will increase to ensure meaningful selection of compound subsets for screening.

Target validation chemistry

Experimental target validation continues to be another major bottleneck in drug discovery research (*7,8*). For informatics disciplines, the key here is to help bridge the gap between the genome, proteome (best defined as the sum of all detectable gene products), and therapeutically relevant protein targets (a relatively small subset of the proteome). Chemistry has begun to significantly impact this process through chemical genomics ("finding all ligands for all targets") and chemical genetics ("using small molecular probes for selectively modulate target function across target families") approaches (*12,13*). Chemoinformatics provides significant opportunities to support these efforts, in particular, through the design, evaluation, and management of target-focused compound libraries or the design of compounds that specifically affect single targets within a family, as very well-illustrated in the study of protein kinases (*13*). Clearly, a deeper understanding of phenomena such as, for example, protein folding, the role of protein flexibility for function, or metabolic pathways (and, ultimately, reliable predictions) would much improve the basis for effective protein target validation, beyond genetic manipulations or chemical interference.

Converging Disciplines

Another question that we feel might be well-worth asking is whether or not it is meaningful to strictly distinguish between bio- and chemoinformatics as disciplines? Clearly, in the context of the popular genome-to-proteome-to-drug paradigm, the distinction becomes at least to some extent artificial, since sequence, structure, and chemical spaces must be "reduced" in a meaningful way to reach the ultimate goal, a new drug candidate (schematically illustrated in Figure 2). This integrated approach is particularly critical during the early stages of discovery. Chemical genomics and genetics (where bioinformatics analysis of target families and chemoinformatics must go hand in hand) provide very good examples for the complementary nature of many of these activities.

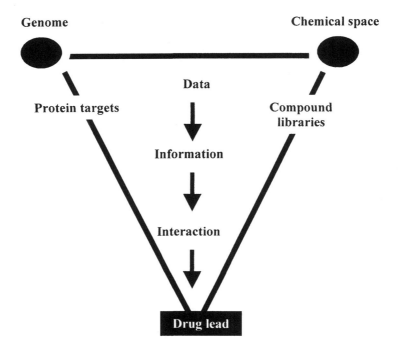

Figure 2. The schematic illustrates the reduction of biological and chemical data and the derivation and exploitation of knowledge about specific interactions (following the "gene-to-proteome-to-drug" or similar paradigms).

In drug discovery settings, such insights have already changed the informatics landscape and triggered the introduction of more global concepts. This is reflected by the increasing use of terms such as "research informatics" or "drug discovery informatics" (*14*). These concepts not only attempt to unify diverse data structures but also to integrate diverse biological and chemical systems that are under investigation (*14*).

Similar Algorithms, Diverse Applications

From a discovery point of view, such integration efforts certainly make sense. But what about the scientific basis for closer integration of (or overlap between) biological and chemical informatics? Figure 3 summarizes what we call the "hierarchy" of biological and chemical informatics, which essentially describes the subjects at different stages of typical bio- or chemoinformatics analysis.

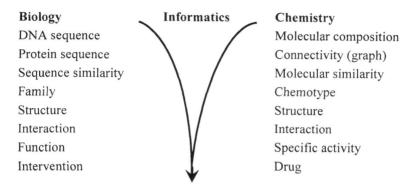

Figure 3. Hierarchical organization of topics in bio- and chemoinformatics

Proceeding from molecular composition to two- and three-dimensional structure and function presents a number of similar challenges, regardless of whether the starting point is DNA or a chemical element distribution (or whether we are focusing on the macromolecular or small molecular level). This general theme has methodological implications. A variety of algorithms and computational techniques that are widely used in chemoinformatics such as clustering methods, principal component analysis, genetic algorithms, self-organizing maps, or neural nets (*3*) are also crucial for many applications in bioinformatics such a, for example, the analysis of microarray data or the correlation of expression patterns and experimental conditions (*15*). Thus, many algorithms and informatics methods are transferable and applicable to a wide range of biological and chemical problems. This supports the design and implementation of more global informatics concepts in life science research. To give an example, the development of advanced relational databases linking test compounds with assay data and other types of biological target-related or pharmacological information provides an important component at the interface between experimental and computational research as well as biology and chemistry. Clearly, as information and knowledge grow, the boundaries between these disciplines begin to disappear.

Summary and Outlook

Much progress has been made in recent years in the development of computational methods for organization and mining of chemical databases or molecular property predictions. As young and rapidly evolving discipline, chemoinformatics still defines itself and we can anticipate that the spectrum of chemical informatics approaches will continue to undergo significant changes for some time to come. To appreciate some of the current activities and trends in this field, it is helpful to consider that much of what we understand as chemoinformatics today has been evolved in drug discovery environments.

Despite the development of many new computational techniques for chemistry applications, some important scientific topics await further progress. For example, the accuracy of computational models and its immediate impact on chemistry programs could be much improved, if it were possible to more accurately predict molecular binding energies or synthetic yields. Importantly, in drug discovery, informatics methods are challenged to address a major bottleneck, the very high attrition rates of clinical candidate molecules. It is therefore not surprising that ADMET parameters are already beginning to be addressed by informatics approaches during the early stages of discovery. However, the computational analysis of these late stage problems is currently still in its infancy.

Other informatics approaches currently aim at the design and compilation of increasingly large databases that combine biological and chemical knowledge such as protein structures, cellular functions, known ligands and binding profiles, or screening data. It is anticipated that such databases will not only substantially grow in years to come, but that their architectures will mature, providing easier and more widely available access. In addition, there are certainly other opportunities for further development and growth of chemically oriented informatics R&D that are not mentioned herein. In general, however, many of these efforts are likely to significantly depend on integration, either with experimental disciplines or bioinformatics.

References

1. Brown, F. K. *Ann. Rep. Med. Chem.* **1998**, *33*, 375-384.
2. Bajorath, J. *Drug Discov. Today* **2001**, *6*, 989-995.
3. Bajorath, J. *J. Chem. Inf. Comput. Sci.* **2001**, *41*, 233-245.
4. Beresford, A. P.; Selick, H. E.; Tarbit, M. H. *Drug Discov. Today* **2002**, *7*, 109-116.
5. Walters, W. P.; Murcko, M. A. *Adv. Drug Deliv. Rev.* **2002**, *54*, 255-271.
6. Pearlman, R. S.; Smith, K. M. *Perspect. Drug Discov. Design* **1998**, *9*, 339-353.
7. Drews, J. *Science* **2000**, *287*, 1960-1964.
8. Stahura, F. L.; Bajorath, J. *Drug Discov. Today* **2002**, *7*, S41-S47.
9. Russo, E. *Naturejobs*, September 12, 2002, p 4.
10. Engels, M. F. M.; Venkatarangan, P. *Curr. Opin Drug Discov. Develop.* **2001**, *4*, 275-283.
11. Bajorath, J. *Nature Rev. Drug Discov.* **2002**, *1*, 882-894.
12. Caron, P. R.; Mullican, M. D.; Mashal, R. D.; Wilson, K. P.; Su, M. S.; Murcko, M. A. *Curr. Opin. Chem. Biol.* **2001**, *5*, 464-470.
13. Shokat, K.; Velecca, M. *Drug Discov. Today* **2002**, *7*, 872-879.
14. Claus, B. L.; Underwood, D. *Drug Discov. Today* **2002**, *7*, 957-966.
15. Butte, A. *Nature Rev. Drug Discov.* **2002**, *1*, 951-960.

Chapter 5

Mathematics as a Basis for Chemistry

G. W. A. Milne

Middle House, Fulbrook, Oxon OX18 4BA, United Kingdom

Mathematics underlies much of chemistry and is used as a tool in many branches of the science. In some areas, particularly those in which structure manipulation is important, mathematics has exerted a more significant pressure and has been pivotal in the development of these chemistry sub-disciplines.

Introduction

It is common to view the sciences in a continuum ranging from the hard, or deductive sciences at one extreme to the soft or descriptive sciences at the other. In such a framework, mathematics clearly occupies one extremity, closely followed by physics. Further along this scale, one finds the more descriptive sciences, such as chemistry and biology and further still, subjects such as medicine, psychology and psychiatry which traditionally, have been highly empirical. A perspective held by many scientists is that this series is, in the intellectual sense, vertical, with mathematics at the top; the harder sciences are in some way the progenitors and disciplines lower down the scale stem from and depend upon these harder sciences for their rationales.

This implication has some validity but it fails to recognize that empirical science is totally legitimate and can and does make progress with little reference to mathematics. It is safe to suggest for example, that botany has advanced to its present state of development with little or no assistance from the hard sciences. It may be less safe however to suppose that botany and mathematics will never benefit from one another. The work of Adleman (*1*), discussed below, in which the DNA molecule is used as a digital computer indeed suggests the opposite.

This paper focuses on the role mathematics has in chemistry, the relationship of the two sciences and the impact this has had upon the *Journal of Chemical Information and Computer Sciences*.

Chemistry Amongst the Sciences

Most chemists would place chemistry near to physics and mathematics in the scale under discussion. This however has not always been the case. Until the 20th century chemistry was a purely empirical science, much as biology is today; its move from the cushions of empiricism to the hard bench of numericism began in the latter part of the 19th century and while it is now well advanced, it has spawned two branches of the discipline. Many chemists view chemical phenomena as potentially predictable and calculable while those more attracted to observational science thrive in, for example, organic synthesis. Both schools make significant contributions to science and society but there is a continuing pressure, from many sources, to convert chemistry from an empirical to a theoretical discipline. This conversion is intellectually attractive because it signifies an understanding of the science. It also has many practical benefits that arise primarily from the control and predictability promised by a theoretical discipline.

Macroscopic and Microscopic Chemistry

Chemistry can be studied at the microscopic or the macroscopic level. At the microscopic level, one is concerned about the details of the structure or behavior of single molecules while at the macroscopic level the focus is on the "real-world" behavior of associations of very large numbers of molecules. These are truly separate camps; a chemist in the first may acquire a profound understanding of the structure of benzene but still be unable to guess at its boiling point, determination or estimation of which are trivial exercises for someone working at the macroscopic level.

Many feel that an adequate understanding of chemical events at the microscopic level will inevitably lead to an unraveling of the problems in macroscopic chemistry. This may be true, but others take the view that macroscopic behavior must be studied for its own sake. For the first group, mathematics is a necessary tool for fundamental research in chemistry while for the second, mathematics is an aid with which observations at the macroscopic level can be more easily recorded and analyzed. These two viewpoints have become the bases for separate sub-disciplines which, respectively, may be termed *mathematical chemistry* and *chemical mathematics*. Mathematical chemistry uses mathematics strictly as a tool and is discussed only briefly here. Chemical mathematics on the other hand has seen much creativity in the approach by mathematicians to long-standing problems in chemistry and is the main subject of this chapter.

Mathematical Chemistry

The credo of the mathematical chemist is that chemical behavior, at any level, should be predictable given a complete understanding of atoms and molecules. In this light, mathematicians were entirely correct is tackling the structure and properties of atoms as a first target. Problems of atomic and molecular structure require solution of the Schrödinger equation which is a non-trivial mathematical exercise. Much of this task has been reduced to commercially available computer code such as *Gaussian* and solutions with considerable precision are now obtainable. In macroscopic problems however, much larger systems must be managed and approximations have been necessary in order to complete the calculations associated with such systems.

Chemical Mathematics

In chemical mathematics, the goal is to use mathematics to restate, explain and rationalize chemical knowledge and open new approaches to the prediction of chemical phenomena. Graph theory has assumed a major role in this process and has facilitated some novel developments in chemical structure theory.

a. Graph Theory

Graph theory is a branch of mathematics which has created a new approach to chemistry and, while it has not revolutionized the science, it has registered a significant impact on it. It has attracted many gifted mathematicians to consider chemical structure and, with the intense activity it has generated in the last 30 years, it has become established as a *bona fide* branch of both mathematics and chemistry.

The starting point in chemical graph theory is that a chemical structure is a graph, which can be manipulated with mathematical techniques. A chemical structure is nothing more than a collection of atoms and bonds; if these are regarded as nodes and edges, the structure is a chemical graph. Thus the chemical notation for methylcyclopropane is paralleled entirely by the corresponding graph, described as an "adjacency matrix":

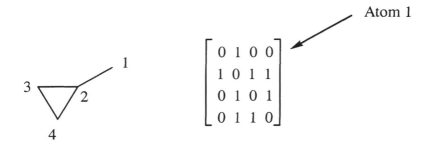

In the matrix, each row describes an atom. Thus in the first row, atom 1 is joined to atom 2 but not to atoms 1, 3 or 4. The second row describes atom 2, bonded to atoms 1, 3 and 4, and so on. All the information in the structure is in the matrix and now can be manipulated mathematically.

In recent years a great deal of effort has gone into exploring the possibilities of this matrix manipulation and it has been shown convincingly that matrices can be constructed in which the data are a full and accurate representation of the structure. Algorithms that search in large databases for specific structures or substructures rest on a basis in graph theory and although the early programs were developed from the chemical perspective, the search algorithms all flow from graph theory. Likewise, graph theoretical representations of chemical structures have proved useful in computer-assisted synthesis design. Enumeration of isomers is a task which can only be done reliably by means of chemical graph theory. The exploration of structure-property relationships is enormously facilitated by graph theory because a graph is a mathematically represented structure and can, by means of statistical techniques, be related to any numeric property of the molecule in question.

These are important areas whose exploration, prior to the application of graph theory, was very incomplete. The mathematical approach has yielded valuable results and each of these topics will be treated very briefly here.

b. Structure Searching

In the 1960s, a group at Stanford headed by Lederberg and Djerassi published a number of papers (2) on what was known as the "Dendral Project". This was a remarkably prescient attempt to deal with problems in organic chemistry from a mathematical viewpoint. The Stanford group explored the problems of structure searching and of isomer enumeration and attempted to identify structures from the corresponding mass spectra. The project, which lasted for over a decade, had limited success because, as is now clear, excessively ambitious problems were tackled.

The basic chemical graph, as can be seen from the example above, contains only carbon and only single bonds. Embellishment of both the structure and the graph gives:

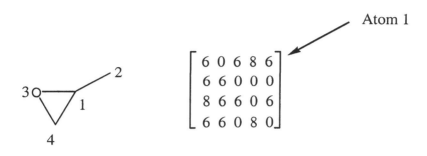

in which atom 1 is clearly identified in column 1 as a carbon (atomic number 6) bonded to two other carbons and to one oxygen (atomic number 8). A matrix like this clearly can be used to search for like fragments in a database. Management of non-carbons and of multiple bonds adds complications and most of the mathematical chemists who worked in this area, chose, after the Dendral experience, to work with alkanes - with only single bonds - and hydrogen atoms are ignored for the sake of simplicity. Such approximations have been the basis of significant criticism from the chemistry camp, but they have permitted the graph theoreticians to make progress with a variety of problems.

A development that was essential for successful structure searching was the Morgan algorithm (*3*), developed at the Chemical Abstracts Service in 1965. This algorithm allowed a canonical and reproducible numbering of the atoms in a molecule which made atom-by-atom comparison of structures possible. Instead of numbering atoms from left to right or top to bottom, Morgan showed that atoms could be numbered hierarchically, from the most highly substituted to the least substituted. Thus Morgan numbering changes the methylcyclopropane example shown above to:

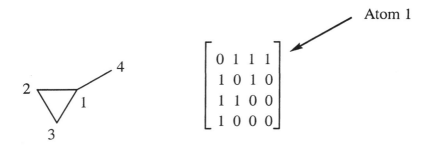

and his algorithm guarantees that this numbering will be used every time this structure is encountered. It should be noted that a simple sort of the rows of the matrix places atom 1, the central atom in the Morganized structure, on the top row.

In 1970, the first widely used algorithms for structure searching became available. In essence, these programs sought to identify each non-hydrogen atom in a structure in terms of its neighbors and then search the database for structures with just such an array of atom-neighbor combinations. Once all such candidates have been found they can be examined one by one to determine if the query structure is indeed imbedded in the retrieved structure. Thus with a query structure 1, the first step will retrieve 2, 3 and 4, but only 4 will survive the atom-by-atom examination. It is this atom-by-atom search for which the Morgan algorithm is crucial.

The early programs were cumbersome because they depended upon the user to define the query structure and to manage much of the complex Boolean logic that was necessary to distill the raw search results to a useful form. In the thirty years since then, many new algorithms have appeared; a new search program now barely rates a publication. The modern programs have taken advantage of the greater compute power that is readily available and they have eliminated all of the troublesome details that characterized their predecessors. It is now commonplace to frame a simple question ("Find this substructure") and elicit a simple answer ("It is contained in 934 structures in this database") but the improvements are entirely in the user interfaces. The basic search process is no different from what it was in the Dendral algorithms.

The underlying graph theoretical principles in these search algorithms have allowed the development of much more sophisticated search techniques. The simple substructure search has always been heavily used because it enables one to answer important questions, but it has now been reinforced by superstructure and similarity searches and by practical methods for the systematic exploration of chemical space(*4*), a matter that has become important recently with the advent of combinatorial chemistry and virtual libraries.

Techniques which, given a query structure, can find all structures in a database which contain it (substructure searching) or which it contains (superstructure searching) are essential techniques in chemical information management. Thus, the indole unit (5) is a substructure of carbazole (6) but a superstructure of pyrrole (7). Both 6 and 7 are "similar" to 5, but in different ways. Any biological activity possessed by 5 is expected to be present also in 6 and so this leads the medicinal

5   **6**   **7**

chemist from a single known active compound to many more compounds with probable activity. Superstructures are important because they begin to define the chemical space in question. If all structures in a database were arranged by chemical "similarity" then 6 and 7 (and 5) would fall near one another and this also carries implications that are important. If 5 resides in a sparely populated chemical environment then this environment (*i.e* any structures related to 5 or 7) are probably worth examining and on the other hand, if the environment is densely populated such research may not be so worthwhile because useful new information is unlikely to result and the possibility that this area has been studied (and patented) by others is considerable.

All of this deals with the type and number of compounds that are chemically similar to the structure of interest. If these are to be likened to neighboring communities then the question arises as to the nature and outer limits of the space, *i.e.* the national or planetary boundaries. This is a far more difficult question which is taken up under Isomer Enumeration, below.

c. Approximations in Chemical Graph Theory

As mentioned above, reduction of chemical structures to their most basic level of fully saturated, acyclic hydrocarbons facilitated the application of graph theory to chemistry. The neglect of cyclic structures in early graph theory work simply meant that a very large segment of organic chemistry, actually about 92% by CAS estimates, was beyond the reach of graph theory. The difficulties of dealing with cyclic structures have since been completely resolved. The second approximation, to ignore heteroatoms, left the graph theory camp open to a major criticism because most chemical behavior results from the presence in a molecule of "heteroatoms", primarily oxygen or nitrogen. Heteroatoms exert an enormous effect upon the distribution in a molecule of electrical charge and it is this that dictates much of the chemical behavior of molecules. Chemical graph theory however blithely ignored charge distribution and so left itself open to very serious criticism from the empirical camp, which knew well how important this is. This difficulty has been addressed in recent years by workers such as Hall and Kier who have developed a valence-connectivity index (*5*) and an "electrotopological" state (*6*),(*7*) which seek to account for atomic charges as well as atomic environment. These methods has since been used with success in a number of applications (*8*),(*9*),(*10*) and have become standard procedures.

It might be noted here that while Hall and Kier's handling of charge in organic graphs has proved itself in many highly successful applications, it is fairly simplistic, particularly in contrast to the very precise calculations of atomic charge allowed by the Schrödinger equation. Here is a gulf, surrounded on both sides by armies eager, but so far unable to traverse it.

d. Isomer Enumeration

Enumeration of isomers has a curious history. Its practical significance is very limited but it poses irresistible intellectual challenges and for at least 250 years scientists have attempted to discover methods for the determination of the number of possible isomers corresponding to a given molecular formula.

The early methods were quite empirical and addressed specific questions such as the numbers of isomeric alkanes and so on. Attempts to generalize the problem began with Cayley in 1874 (*11*) and were developed further by Henze and Blair (*12*) and by Pólya (*13*). With the availability of computers a renewed effort to systematize these problems was undertaken; in 1981, Knop and coworkers (*14*) published an algorithm which correctly enumerates the number of isomeric acyclic alkanes for a given carbon number.

It is of interest here not to pursue the details of isomer enumeration but to contemplate some of the results, which have a bearing on the concept of chemical space, referred to above. When considering a specific chemical structure, primary questions concern the number of structures which contain it as a substructure and the population of compounds which are themselves substructures of the species in question. Looming over these however is the larger question as to how many structures exist in chemical space in general. Isomer enumeration provides some partial answers to this. If one limits the elements in organic compounds to the "organic elements" C, H, N, O, S, F, Cl, Br and I, then in principle, it is straightforward to generate all possible molecular formulas and then to calculate the number of possible structures corresponding to each molecular formula. The total number of structures generated in this way defines the chemical space occupied by compounds composed of these elements.

Such calculations have never been completed, but from the work that has been done, it is quite clear that very large numbers are involved. A figure of 10^{20} - 10^{24} has been proposed by Ertl (15). This is only an estimate (note the range of 5 orders of magnitude) but the current size of the CAS Registry, which is 2 x 10^7 compounds, gives one the sense that 10^{24} is not totally unreasonable. Isomer enumeration calculations have shown[18] for example, that there are 1,117,743,651,746,953,270 (*i.e.* about 10^{18}) different isomers of $C_{50}H_{102}$ and 40 times as many isomers of $C_{50}H_{102}O$. These numbers cannot include any cyclic or unsaturated compounds, nor of course any compounds containing elements other than C, H and O. The number 10^{18} is only 10^{-6}% of total chemical space; whether or not this is plausible is arguable but it seems fair to say that with only 2 x 10^7 compounds currently recorded in the CAS Registry, there are very many more chemicals waiting to be prepared and characterized. It also is perhaps worth noting that combinatorial chemical syntheses, which when particularly adventurous, might examine a million structures, are barely scratching the surface of what is possible.

e. Topological Indices

Once a chemical structure has been converted to an equivalent matrix, it is possible to manipulate this matrix in a variety of ways to generate a single number which describes the molecule more or less completely. These molecular descriptors are usually known as a *topological indices* and have been shown to possess a number of useful applications. If the calculated index is non-degenerate, associated with only one structure, it becomes a useful identifier and because of this, the uniqueness of topological indices is an important characteristic. Some indexes, such as Balaban's J indices, are highly non-degenerate but the belief is that absolute non-degeneracy can never be achieved and that as molecular properties coincide then so will topological indices. A more important practical application stems from the observation that topological indices reflect molecular shape in some way and often track and can be regressed to physical properties such as boiling point which themselves are related to molecular shape. Once a successful regression is in hand, the regression equation can be used to predict the properties of new structures.

One of the earliest and simplest topological indices is the **Wiener Index** (*16*) which is defined as the sum of the number of bonds between any pair of carbon atoms. Thus for isopentane:

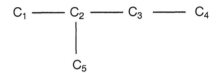

the following table can be constructed:

Atom pair	Number of bonds
1,2	1
1,3	2
1,4	3
1,5	2
2,3	1
2,4	2
2,5	1
3,4	1
3,5	2
4,5	3

and the number of bonds in the structure, the sum of the right-hand column in the table is thus 18. This is the Wiener Index (w) for isopentane. An easier way of performing this calculation is to consider each C-C bond in turn and multiply the number of carbons on one side of the bond by the number on the other. For isopentane, one gets:

$$4*1 + 2*3 + 1*4 + 1*4 = 18$$

In the same way, the number of pairs of carbons (p) separated by 3 bonds can be determined. In this case there are only 2 such pairs (C_1-C_4 and C_4-C_5) and so for isobutane, w = 18 and p = 2. Wiener showed that the variable w and p can, by least squares regression, be related to the boiling point of the compound by the relationship:

$$bp = aw + bp + c$$

This finding, that a physical property can be predicted from the structure of the compound, precipitated a flood of research projects aimed at broadening and exploiting it.

The **Hosoya Index**(*17*) attempted to characterize the overall topology of a molecule by summing the numbers of structural patterns with one, two, three... disjoint bonds. As an example, 2,3-dimethylhexane has 7 single disjoint bonds, 13

double disjoint bonds and 6 triple disjoint bonds, and no quadruple or higher

1 disjoint bond, $Z_1 = 7$

2 disjoint bonds, $Z_2 = 13$

3 disjoint bonds, $Z_3 = 6$

disjoint bonds, giving a Hosoya Index of $1 + 7 + 13 + 6 = 27$. The Hosoya Index and Indices derived from it regress very well to the boiling point data for acyclic hydrocarbons, giving the equation:

$$BP = 1.2211Z + 81.6939$$

which predicts boiling points with a standard error of $2.90°$, comparable to experimental error.

Randic (18) has developed a "**Connectivity index**" which seeks to properly evaluate the effect on a structure of branching. Each bond in a structure makes a contribution of $1/\sqrt{m.n}$ to the connectivity index ($^1\chi$) where m and n are the number of carbons attached directly to the carbons involved in the bond. Thus for the 1-2 bond in 2,3-dimethylhexane, atom 1 has 1 neighboring carbon and atom 2 has 3, thus $m = 1$ and $n = 2$. Likewise for the 2,3 bond $m = 3$ and $n = 3$ and so on. For the whole molecule;

$$^1\chi = 1/2 + 3/\sqrt{3} + 1/3 + 1/\sqrt{6} + 1/\sqrt{2} = 3.6807$$

The connectivity index can be calculated for all acyclic hydrocarbons; some selected values for the isomeric dimethylhexanes are given in the table below.

Compound	1χ
2,2-Dimethylhexane	3.5607
3,3-Dimethylhexane	3.6213
2,5-Dimethylhexane	3.6259
2,4-Dimethylhexane	3.6639
2,3-Dimethylhexane	3.6807

The connectivity index and its derivative, the "**ID Number index**" both give special weight to the peripheral bonds 1-2, 2-7 etc. and less weight to the "buried"bonds 2-3, 3-4, and 4-5. Thus the contributions of different bonds to the overall molecular surface are expressed in this index.

At least 200 different topological indices has been proposed during the last decade and in fact in the view of many, they have begun to clutter the literature. These topological indices have been used a great deal in QSAR and QSPR studies.

For any such index, its non-degeneracy and the quality of the regressions it supports are important properties and these have both been examined in some detail by Randić (*19*).

Just as different snapshots of a person can emphasize or de-emphasize different features, so different topological indices can accentuate different structural features such as branching. Thus the Wiener index w may be thought of as an expression of the mean molecular volume and its cube root $\sqrt[3]{w}$, the diameter(*20*). Likewise, the connectivity index[22] reflects branching in a structure.

Finally, topological indices can be regarded as a measure of similarity and this has been used in some algorithms that attempt to calculate structural similarity.

f. Reverse Processing

Topological indices have one important drawback; they are calculated in a process that is not reversible. Thus to derive any particular topological index from a structure is routine but the reverse process has never been accomplished and so, though a valid analysis may provide an equation relating a physical property to a topological index, there is no general method to use the topological index to derive a corresponding structure. In an effort to address this problem, Zefirov et al. developed(*21*) a regression equation relating the Hall and Kier "Kappa Indices(*22*)" κ_1 (size), κ_2 (shape) and κ_3 (centrality) to the heat of evaporation:

$$\Delta H_e = 3.971\kappa_1 + 1.285\kappa_2 - 0.253\kappa_3 + 2.683$$

This equation was them solved to find all the C_6 alkanes with a heat of evaporation between 28 and 30 kJ/mol. The three compounds in the table below were identified.

Compound	ΔH_e (kJ/mol)
2-Methylpentane	29.86
3-Methylpentane	30.27
2,3-Dimethylbutane	29.12

g. Genetic Algorithms

Beginning in the early 1990s, several hundred papers appeared on genetic algorithms and this has now been established as a method for the prediction of the properties of chemicals. An application of a genetic algorithm in polymer design was reported by Venkatasubramanian et al.(*23*) in which the goal was to design a polymeric structure with specific physical properties. This method depends upon the ability of an algorithm to generate a family of structures in a manner which is powered by genetic selection and guided by the required physical property. It is summarized in the flowchart below.

A seed population of typical polymer structures is used as a starting point. The

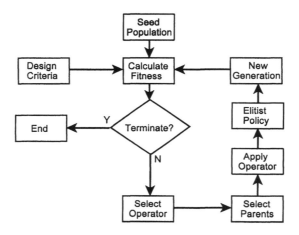

"fitness" of each structure (the inverse of the similarity of the estimated property and the desired property, a normalized number which should approach 1) is calculated with reference to the design criteria then, with a genetic operator in place, two "parents" are selected and an offspring structure is produced by crossover or mutation. If the fitness of the offspring is improved over that of the parents the offspring is selected and processed further. External "elitist factors" such as stability and environmental acceptability can be incorporated into the decision making process. In the case reported, polymers were designed with properties (density, glass transition temperature, coefficient
of thermal expansion, specific heat and bulk modulus) that fell within 0.6% of the desired values.

This algorithm allows the design of structures with a specific property and is of major economic importance. Algorithms of this sort are now in use in a variety of industries.

h. Statistics

A great deal of chemistry is susceptible to statistical analysis but in spite of this, statistics is not typically included in the college chemistry curriculum. One outcome of this is that there are many qualified chemists attempting statistical analysis of chemical problems and in the process, committing grievous errors. This, as has been pointed out(24), is a "dangerous but entirely curable affliction".

Statistical mechanics - the probability of different phenomena in an assembly of a large number of molecules and uses - is a well-established discipline and will not be considered in detail here. An exercise of statistics that is more relevant in this context concerns the analysis of a particular property, such as toxicity, over a large number of different structures. This type of analysis is known as a Quantitative Structure Activity Relationship (QSAR) or a Quantitative Structure Property Relationship (QSPR) study and there are thousands of examples of these in the literature. During the 1970s, a considerable amount of work was done in the pharmaceutical industry to relate properties such as the octanol/water partition coefficients of different molecules to their biological activities. These enjoyed some success but because properties such as solubility were themselves unpredictable, the practical value of these studies was limited. Many of the properties used in such studies are in effect, surrogates for the structure of the compound, and the relationship that is most useful is the direct one, between structure and property.

A topological index (TI) is an alternative expression of the structure and if it can be shown that the TI is statistically related to a property such a specific bioactivity, then such a relationship will be valuable because it will limit the amount of biological testing necessary to discover a lead drug. In principle, it should also allow the identification of the optimum structure for a given property, but here there is a barrier in the form of the "reverse problem", discussed above. One may well discover the optimum value of a TI but finding the structure that corresponds to this TI is difficult.

In work carried out in the 1980s, relationships relating vapor pressure to experimentally determined properties such as boiling point, critical pressure and so on had been reported but in 1998, Liang and Gallagher showed(25) that vapor pressure p_L could be calculated as a function of the polarizability of the molecule α and the counts in the structure of hydroxyl, carbonyl, amino, carboxylic, nitro and nitrile groups:

$$\log p_L = -0.432\alpha - 1.382(OH) - 0.482(C=O) - 0.416(NH) - 2.197(COOH) - 1.383(NO2) - 1.101C\equiv N + 4.610$$

In 2002, Quigley and Naughton(26) showed that the effectivenenss of β-blocking agents, expressed as the angor treatment dose (ATD) is related to the connectivity index $^1\chi$ and the eccentricity index $\xi^{A(27)}$ by the equation:

$$ATD = (193.70 \pm 63.74)\ ^1\chi - (85.38 \pm 58.9)\ \xi^A - (798.55 \pm 702.4)$$

a correlation for which $R^2 = 0.781$ and $s = 174.7$. The acute toxicity, LD_{50} is given by the relation ($R^2 = 0.781$) established by Pérez-Giménez et al(28):

$$LD_{50} = (7.742 \pm 1.721)\ ^1\chi^v - (2.266 \pm 1.605)\ \xi^A - (49.876 \pm 12.21)$$

or by the slightly better ($R^2 = 0.933$) relationship derived by the Gálvez group(29)

$$LD_{50} = (0.158 \pm 0.075)\ ^1\chi^v - (0.268 \pm 0.060)\ \xi^A - (3.324 \pm 0.650)$$

Basak's group reported in 2000(30) that the acute aquatic toxicity of series of benzene derivatives could be determined by regression using topostructural indices, topochemical indices, geometric parameters and quantum chemical parameters. The best correlation obtained used seven parameters:

$^0\chi$	path connectivity index, order 0
P_9	number of paths of length 9
IC	information content of the distance matrix partitioned by the frequency of occurrence of distance h
$^5\chi^v$	valence path connectivity index, order 5
$^{3D}W_H$	3D Wiener Number for hydrogen-filled geometric distance matrix
Δhf	heat of formation
μ	dipole moment

and predicted the LC_{50} of the compound with $R^2 = 0.861$ and $s = 0.30$.

As can be seen from this brief survey of some examples it is important, as a practical matter, that estimation of properties be accomplished without reference to measured quantities such as boiling point. The work of the β-blockers meets this requirement but Basak's aquatic toxicity work, which uses heat of formation and dipole moment, does not. It is also clear that the use of topological indices greatly assists the search for correlations between theoretical parameters and actual properties of molecules and this work should, and surely will continue.

i. DNA Computing

The use of DNA as a computing device was touched on earlier, and is an important innovation, which deserves a fuller treatment. The natural role of DNA requires, amongst other things, a computational ability and it is thought that DNA sequences might be usable in the encoding of information for mathematical systems.

In the work of Adelman[1] which is regarded as seminal, a solution to a seven city variation of the "traveling salesman" problem was developed using DNA as the computer.

The traveling salesman problem, also known as the Hamiltonian path problem, is a classical problem in graph theory. It is to find the shortest path which visits each of n cities just once each as shown in the diagram below, in which n = 7.

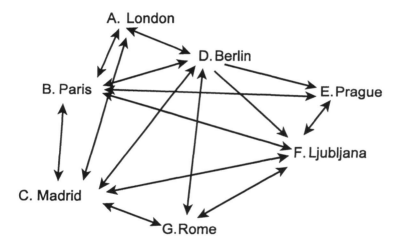

This is an "NP" (non-deterministic polynomial time) problem and, once n > 20 it can only be solved by massively parallel computers because it is combinatorially explosive. An NP problem such as this must generate random paths through this graph, keep only those that begin in London and end in Rome, and keep only those that visit all seven cities once each. This can be solved with a DNA computer as follows (*31*):

For each of the cities A - G, a random sequence 20-mer oligonucleotide is made. The complementary strands A' - G' are also made. For every path A → B,

B →C, and so on, a 20-mer is made consisting of the 3' 10-mer of the originating city and the 5' 10-mer of the destination. Using A and G' as primers, this mixture is subjected to ligation and PCR amplification which will create many multi-city

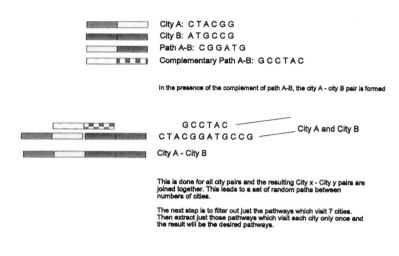

The Traveling Salesman Problem

starting at A and ending at G. The mixture obtained is passed through an agarose gel and all but the 140 bp band (paths through 7 cities) are discarded. These 7-city paths are then checked to ensure that they all go through each of the original cities. The resulting oligomers:

a. Cover 7 cities
b. Begin at London (A) and end at Rome (G)
c. Include every city and therefore include no city more than once

These are thus the possible solutions to the Traveling Salesman problem. Analysis of the mixture and sequencing of the products gives the full result. Calculation of the total distance associated with each path is simply arithmetic.

It is important to note that this is an NP problem, which cannot in general be solved by sequentially operating computers. Massively parallel computers can handle such problems and a DNA computer is a massively parallel machine. It handles each calculation relatively slowly but can manage millions of parallel problems simultaneously.

Afterword

This paper has attempted to examine the impact of mathematics upon chemistry. The question is: Has mathematics exerted a fundamental force in chemistry or has it served merely as a tool?

The answer is, with one major exception, mathematics has served only as a tool in chemistry. The major exception of course is graph theory. Graph theory has provided and new, sometimes superior way in which chemical structure can be viewed. The results from chemical graph theory are very important; an ability to reliably and accurately predict the properties of chemical structures could revolutionize virtually the entire chemical industry. The impact in the petrochemical industry is already clear and such methods, once available, will certainly be adopted in other areas, such as the detergent, agrochemical and pharmaceutical industries, whose major expenses are related to synthesis and testing of chemical compounds.

Elsewhere in chemistry, as has been seen, mathematics is used heavily and frequently permits the solution of otherwise intractable chemical problems.

References

1. Adleman, L. Molecular computation of solutions to combinatorial problems. Science 266, 1021-1024, (1994).

2. See, for example, Lederberg, J., Sutherland, G. L., Buchanan, B. G., Feigenbaum, E. A., Robertson, A. V., Duffield, A. M., Djerassi, C. Applications of Artificial Intelligence for Chemical Inference. I. The Number of Possible Organic Compounds. Acyclic Structures Containing C, H, 0, and N. J. Amer. Chem. Soc., 91, 2973-2976, (1969) and references cited therein.

3. Morgan, H. The Generation of a Unique Machine Description for Chemical Structures - A Technique Developed at Chemical Abstracts Service. J. Chem Doc, 1965, 5, 107-113. See also: http://www.xtal.iqfr.csic.es/isostar/help/morgan.html

4. Faulon, J.-L. Stochastic Generator of Chemical Structure. 2. Using Simulated Annealing to Search the Space of Constitutional Isomers. *J. Chem. Inf. Comput. Sci.*, **36**, 731-740, (1996).

5. Kier, L. B., Hall, L. H. Molecular Connectivity. VII. Specific Treatment of Heteroatoms. *J. Pharm. Sci.,* **1976,** 65, 1806-1809.

6. Kier, L. B., Hall, L. H. Molecular Structure Description: The Electrotopological State. Academic Press (1999).

7. Hall, L. H., Kier, L. B. Electrotopological State Indices for Atom Types: A Novel Combination of Electronic, Topological and Valence State Information. *J. Chem. Inf. Comput. Sci.,* 1995, **35,** 1039-1045.

8. Huuskonen, J. QSAR Modeling with the Electrotopological State: TIBO Derivative. *J. Chem. Inf. Comput. Sci.,* **2001,** *41,* 425-429.

9. Gough, J. D., Hall, L. H. Modeling Antileukemic Activity of Carboquinones with Electrotopological State and Chi Indices. *J. Chem. Inf. Comput. Sci.,* **1999,** *39,* 356-361.

10. Rose, K., Hall, L. H., Kier, L. B. Modeling Blood-Brain Barrier Partitioning Using the Electrotopological State. *J. Chem. Inf. Comput. Sci.,* **2002,** *42,* 651-666.

11. Cayley, A. *Ber.,* **8,** 1056- (1875).

12. Henze, H. R., Blair, C. M. The Number of Structurally Isomeric Alcohols of the Methanol Series. *J. Amer. Chem. Soc.,* **53,** 3042-3046, (1931); The Number of Structural Isomers of the More Important Types of Aliphatic Compounds. *J. Amer. Chem. Soc.,* **56,** 157-157, (1934).

13. Pólya, G. *Acta Math.,* **68,** 145, (1937).

14. Knop, J. V., Müller, W. R., Szymanski, K., Trinajstić, N. *J. Chem. Inf. Comput. Sci.,* **21,** 94, (1981).

15. Ertl, P. Cheminformatics Analysis of Organic Substituents: Identification of the Most Common Substituents, Calculation of Substituent Properties, and Automatic Identification of Drug-like Biosteric Groups. *J. Chem. Inf. Comput. Sci.,* **2003,** *43;* viewed on the Web as an ASAP Article.

16. Wiener, H. Structural Determination of Paraffin Boiling Points. *J. Amer. Chem. Soc.,* **1947,** *69,* 17-20.

17. Hosoya, H. Topological Index. A Newly Proposed Quantity Characterizing the Topological Nature of Structural Isomers of Saturated Hydrocarbons. *Bull. Chem. Soc. Jpn.* **1971,** *44,* 2332-2339.

18. Randić, M. On Characterization of Molecular Branching. *J. Amer. Chem. Soc.*, **1975**, *97*, 6609-6615.

19. Randić, M., Balaban, A. T., Basak, S. C. On Structural Interpretation of Several Distance-Related Topological Indices. *J. Chem. Inf. Comput. Sci.*, **2001**, *41*, 593-601.

20. Platt, J. R. Prediction of Isomeric Differences in Paraffin Properties. *J. Chem. Phys.*, **1952**, *56*, 328-336.

21. Skvortsova, M. I., Baskin, I. I., Slovokhotova, O. L., Palyulin, P. A., Zefirov, N. S. Inverse Problem in QSAR/QSPR Studies for the Case of Topological Indices Characterizing Molecular Shape (Kier Indices). *J. Chem. Inf. Comput. Sci.*, **1993**, *33*, 630-634.

22. Kier, L. B. Indexes of Molecular Shape from Chemical Graphs. *Med. Res. Rev.*, **1987,** *7*, 417-440.

23. Venkatasubramnaian, V., Chan, K., Caruthers, J. M. Evolutionary Design of Molecules with Desired Properties Using the Genetic Algorithm. *J. Chem. Inf. Comput. Sci.*, **1995**, *35*, 188-195.

24. Miller, J. C., Miller, J. N. Statistics for Analytical Chemistry. 2nd. Edition. Ellis Horwood (1992). ISBN 0-13-845421-3.

25. Liang, C., Gallagher, D. A. QSPR Prediction of Vapor Pressure from Solely Theoretically-Derived Descriptors. *J. Chem. Inf. Comput. Sci.*, **1998,** *38*, 321-324.

26. Quigley, J. M., Naughton, S. M. The Interrelation of Physicochemical Parameters and Topological Descriptors for a Series of β-Blocking Agents. *J. Chem. Inf. Comput. Sci.*, **2002**, *42*, 976-982.

27. Gupta, S., Singh, M., Madan, A. K. Predicting Anti-HIV Activity: Computational Approach Using a Novel Topological Descriptor. *J. Comp.-Aided Mol. Design*, **2001**, *15*, 671-678.

28. Garcia-March, F. J., Cercós-del-Pozo, F., Pérez-Giménez, F., Salabert Salvador, M. T., Jaén Oltra, Antón Fos, G. M. Correlation of Pharmacological Properties of a Group of β-Blocker Agents by Molecular Topology. *J. Pharm. Pharmacol.* **1995**, *47,* 232-236.

29. García-Domenech, R., de Gregorio Alapont, C., de Julián-Ortiz, J., Gálvez, J. Molecular Connectivity to Find β-Blockers with Low Toxicity. *Bioorg. Med. Chem. Lett.,* **1997**, *7*, 567-572.

30. Basak, S. C., Grunwald, G. D., Gute, B. D., Balasubramanian, K., Optiz, D. Use of Statistical and Neural Net Approaches in Predicting Toxicity of Chemicals. *J. Chem. Inf. Comput. Sci.*, **2000**, *40*, 885-890.

31. Will Ryu: http://www.arstechnica.com/reviews/2q00/dna/dna-1.html

Chapter 6

On the Magnitudes of Coefficient Values in the Calculation of Chemical Similarity and Dissimilarity

John D. Holliday, Naomie Salim, and Peter Willett

Department of Information Studies and Krebs Institute of Biomolecular Research, University of Sheffield, Western Bank, Sheffield S10 2TN, United Kingdom

Analysis of the distributions of inter-molecular similarity values has been carried out using the Tanimoto coefficient, the Cosine coefficient and the complement of Euclidean distance. In order to determine if they are an effective measure for dissimilarity-based methods, their characteristics at low values have been compared with distributions derived using bit-strings generated by random techniques. The effectiveness of similarity measures for property prediction across the full range of ranked search output was then examined. The results show that the distributions of inter-molecular similarity measures are not random in nature, but their effectiveness for property prediction is better than random only when very small or very large similarity values are considered.

Keywords: Bit-string, Cosine coefficient, Euclidean distance, Fingerprint, Neighbourhood principle, Similar property principle, Similarity measure, Tanimoto coefficient

INTRODUCTION

The measurement of structural similarity plays an important role in many aspects of chemoinformatics, such as database searching (*1*), the design of combinatorial libraries (*2*) and the prediction of biological activity (*3*). While many different similarity measures have been reported in the literature (*4-6*), most applications use a particularly simple approach based on the number of 2D fragment substructures common to a pair of molecules. Such measures were first reported by Adamson and Bush (*7*) but became more widely adopted following studies of database searching at Lederle (*8*) and at Pfizer (*9*) in the mid-Eighties. Current approaches to 2D similarity searching typically use bit-string or fingerprint representations of molecules with the similarity being calculated by the Tanimoto Coefficient (*1*), and systems using this are included in much software for chemical information management.

Given two molecules A and B represented by bit-strings containing a and b non-zero bits, c of which are in common, then the Tanimoto coefficient is defined to be

$$Tan = \frac{c}{a+b-c}.$$

The reader should note that the Tanimoto Coefficient can also be used with non-binary molecular representations, such as a set of calculated physicochemical properties, but we restrict ourselves here to binary representations. The Tanimoto coefficient, which is also known as the Jaccard Coefficient, has values in the range zero to unity, these corresponding to bit-strings having no bits at all in common or having all bits in common, respectively. The complement of the Tanimoto coefficient – also known as the Soergel coefficient – can be used as a dissimilarity coefficient in molecular diversity studies.

Although widely used for similarity searching and diversity selection, the binary form of the Tanimoto Coefficient has occasioned much recent criticism (*10-14*). Godden *et al.* (*12*), for example, showed that this coefficient has an inherent bias to certain specific similarity values, with the highest peak in the distribution of possible similarity values occurring at around 0.3 to 0.4. Since this range of statistically-preferred Tanimoto values is not outside the range of values that might be encountered in diversity-based selection studies (*2, 15*), it is possible that dissimilarity-based compound selection (DBCS), for example, could be influenced by such chance occurrences.

Flower (*11*) supports Lajiness's comment (*10*) that the distribution of Tanimoto coefficients tends to shift towards larger values when comparing more complex query compounds. This is due to the fact that, in a similarity search using fragment bit-strings or fingerprints, a large molecule in the database is *a priori* much more likely to have bits in common with the target structure than a small molecule. Thus, if an association coefficient like Tanimoto coefficient is used in DBCS, small molecules tend to be selected because they are likely to have few bits set in a fingerprint; similar studies have been carried out by Dixon and Koehler (*13*) who have also shown that the Euclidean distance, in contrast, puts greater emphasis on large compounds when selecting a subset of diverse structures. Studies such as these have led to the suggestion that the Tanimoto Coefficient should be modified (*14, 16*) or used in combination with other coefficients (*13, 17*). In this paper, we consider the use of the Tanimoto coefficient for the calculation of molecular similarity and dissimilarity using fragment bit-strings, comparing the results obtained with those from two other common similarity coefficients, the Cosine coefficient and the Euclidean distance (*1*).

LOW-VALUED TANIMOTO COEFFICIENTS

Applications of similarity measures normally focus on high-valued coefficients. For example, the Jarvis-Patrick clustering method requires the initial identification of the *nearest neighbours*, i.e., the most similar molecules, for each molecule in a database as precursor to the actual clustering phase. Again, in similarity searching or property prediction, one may require all molecules having a value for the similarity coefficient with the target structure in the search that is greater than some threshold value, e.g., the use of 0.7 - 0.8 as a Tanimoto cut-off in property prediction (*3*). Conversely, DBCS methods (*10, 18, 19*) involve selecting a (near-) maximally diverse subset of a file of molecules by focusing on the *furthest neighbours*, i.e., those molecules that have the smallest coefficient values, or those that have a coefficient value less than some threshold. Algorithms of this sort are readily implemented using coefficients like the Tanimoto coefficient, but one can reasonably query the appropriateness of such procedures. The problem is illustrated in Figure 1 which shows the target structure for a similarity search of the NCI AIDS database (*20*) using UNITY 2D fragment bit-strings (*21*) together with some of the molecules retrieved in the first decile of the sorted ranking. Visual inspection of Figure 1(b) suggests that there is a clear (topological) resemblance between these molecules and the target structure; moreover, the similarity of the first molecule does appear (to the authors of this paper at least) to be greater than the similarity of the second molecule, in accordance with the ordering suggested by the calculated values of the Tanimoto coefficient. Now consider the structures shown in Figure 1(c), which come from the tenth decile of the sorted ranking. There does not seem to be any obvious resemblance between these molecules and the target structure, and it is not at all clear (again to these authors) that the first molecule is more dissimilar from the target structure than is the second, whereas the Tanimoto coefficient would order the two molecules in the order shown.

Examples such as those shown in Figure 1 suggest that while the Tanimoto coefficient is an appropriate tool for ranking a set of molecules and discriminating between them when large coefficient values are involved, the coefficient appears to be unable to discriminate appropriately when low coefficient values are involved. Yet it is precisely the latter class of values that are the focus of DBCS algorithms.

DISTRIBUTION OF SIMILARITY VALUES

Generation of fragment bit-strings

The discussion in the previous section suggests that it may be appropriate to consider the extent to which Tanimoto values do, in fact, encode meaningful information about the structural relationships between molecules. If we wish to address this question then we must have some external basis of comparison as similarity is, after all, an inherently subjective concept. The approach taken here is to compare the similarity values obtained from sets of real fragment bit-strings with the values obtained from sets of random fragment bit-strings. This approach is analogous to the sequence randomization procedures that are routinely employed to validate the significance of the homology scores obtained in searches of databases of proteins and of nucleic acids; the use of randomised structural representations in the calculation of chemical similarities was first suggested by Bradshaw (*22*) and by Sheridan and Miller (*23*).

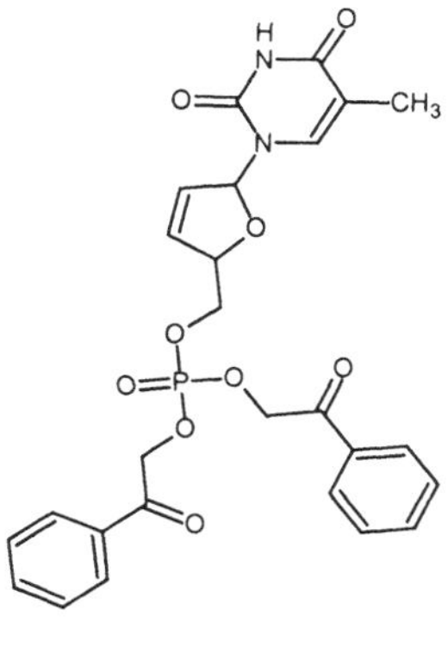

(a) Target structure.

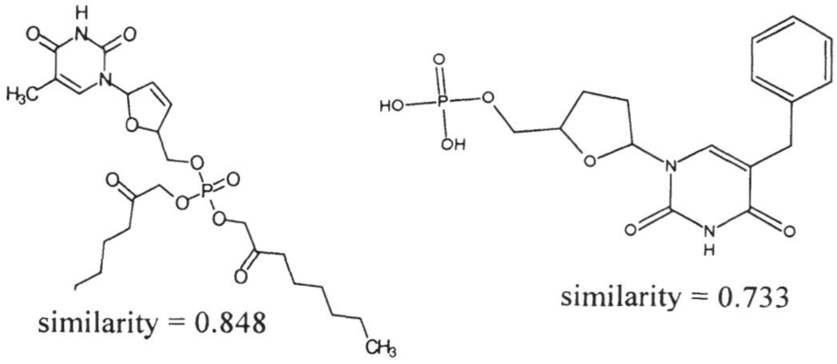

(b) Sample structures from the first decile of sorted ranking.

similarity = 0.848

similarity = 0.733

(c) Sample structures from the tenth decile of sorted ranking.

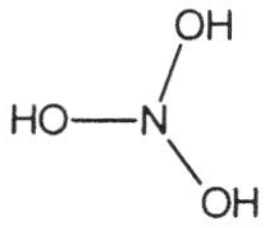

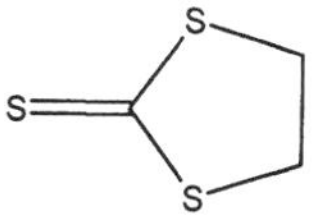

similarity = 0.033

similarity = 0.040

Figure 1. Examples of results from a similarity search against a target using 2D bit string based Tanimoto coefficient. Compounds are from the AIDS database, characterised using UNITY 2D bit strings.

The use of a randomization approach requires some way of generating random bit-strings, so that their behaviour in similarity analyses can be compared with the behaviour of real bit-strings. Here, we have used four different approaches. The first is simply to generate a bit-string containing m non-zero bits by generating m random integers in the range 1-n, where n is the total number of bits in the bit-string. While certainly random, such a bit-string is unlikely to be anything like a real one and we have hence tested three other bit-string generation methods that seek to model real bit-strings more closely. The second approach involved an analysis of the frequencies with which bits were set in the real bit-strings describing a dataset. Then, during the generation process, bits were set with probabilities based on their frequencies of occurrence in these real bit-strings. It is known that there are strong correlations between the frequencies with which fragments occur in sets of molecules (24, 25), and the third set of random bit-strings were hence generated so as to reflect the inter-bit co-occurrences in the real bit-strings. The fourth set of random bit-strings was generated taking account of both the frequencies and the co-occurrence frequencies in the real bit-strings.

Two real datasets were used: the first was a set of 5772 molecules from the NCI AIDS file mentioned previously; and the second was a set of 11607 molecules from the ID Alert database (26). Both of these sets of molecules were characterised by three types of real bit-string. The Barnard Chemical Information (BCI) (27) bit-string is a 1052-bit structural key-based bit-string derived from BCI's standard 1052 fragment dictionary and encoding augmented atoms, atom sequences, atom pairs, ring component descriptors and ring fusion descriptors. The Daylight bit-string (28) is a 2048-bit hashed fingerprint that encodes each atom type, augmented atoms, and sequences of length 2-7 atoms. The UNITY bit-string (21) is a 992-bit fingerprint which encodes sequences of length 2-6 atoms as well as structural keys for common atom and bond types; the bit-string is part-hashed in that it hashes information from different path lengths separately.

Analysis of similarities

Inter-molecular similarities were calculated using three different coefficients for all five bit-strings (the real bit-string and the four random bit-strings) for each of the two databases using each of the three bit-string representations. The three coefficients used were the Tanimoto coefficient, the Cosine coefficient, and the complement of the Euclidean Distance after normalization by n, the number of bits in the bit-string.

Inspection of the frequency distributions demonstrates clearly that the real bit-strings give distributions that are different from the random bit-strings, with the former distributions consistently containing notably larger numbers of very high and very low similarity coefficient values than do the latter distributions. For example, Figure 2 shows the frequency distributions of Tanimoto similarity values for all the molecules containing 60-79 bits, 130-149 bits, and 160-179 bits respectively in their real bit-strings, together with the random results obtained by averaging over ten different random bit-strings in each case. The figure shows the distributions obtained with the BCI bit-strings for the AIDS dataset, but entirely comparable plots are observed with the Unity and Daylight bit-strings, and with the ID Alert dataset, and also for the Cosine and Euclidean Distance coefficients. Use of χ^2 and Kolmogorov-Smirnov tests show that all these are significantly different at the 0.05 level of statistical significance.

These experiments hence demonstrate that the real bit-strings give distributions that are significantly different from those that are obtained from bit-strings generated by random means; moreover, this finding applies throughout the full range of similarity values and using all of the types of bit-string and similarity coefficient that were tested. It hence seems possible to reject the possibility that low-valued similarity coefficients contain no more meaningful structural information than do coefficients generated by random processes.

Analysis of coefficients

The experiments above do provide some support for the continuing use of bit-string coefficients for selecting sets of compounds. However, as noted previously, earlier studies have highlighted several characteristics of similarity coefficients that affect their use in molecular diversity analyses such as DBCS (*10, 11, 13*). These characteristics apply not just to real bit-strings but also to the simulated ones studied here, and are highlighted by our use of three different coefficients, as is exemplified by inspection of Figure 3. Here, the Tanimoto distribution is towards the left-hand end of the similarity axis when compared to the Cosine distribution which covers by far the widest range of similarity values; and the distribution for the complement of the Euclidean distance is strongly peaked and covers only a small range of similarity values. Another characteristic of these three coefficients is the effect of variations in the numbers of bits set on the similarity values that are calculated. Specifically, the Tanimoto and Cosine values tend to increase as more bits are set in the bit-strings that are being compared, while the Euclidean complement values tend to decline as more bits are set. It is quite easy to rationalize this behaviour.

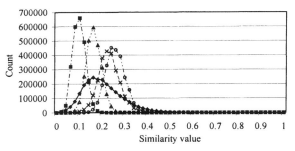

(a) Bit strings with 60-79 bits set.

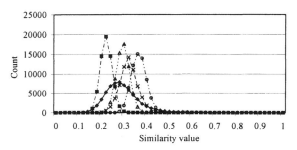

(b) Bit strings with 130-149 bits set.

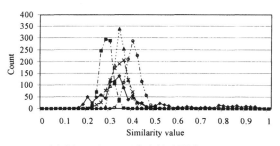

(c) Bit strings with 160-179 bits set.

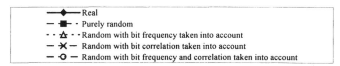

Figure 2. Comparison of Tanimoto frequency distributions of intermolecular similarity values from the AIDS dataset characterised by the BCI bit strings with four randomly generated bit strings.

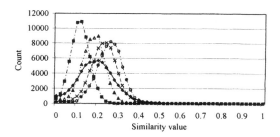

(a) The Tanimoto measure.

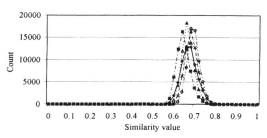

(b) The Euclidean measure.

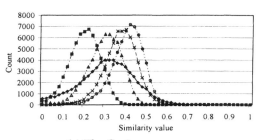

(c) The Cosine measure.

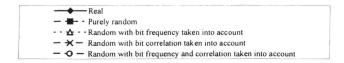

Figure 3. Comparison of Tanimoto, Euclidean and Cosine frequency distributions of intermolecular similarity values.

The Euclidean complement coefficient used in our experiments can be defined as

$$Euc = 1 - \sqrt{\frac{a+b-2c}{N}}$$

where a, b and c have been defined in the Introduction to the paper and where N is the total number of bits in a bit-string. If we assume that $a > b$ then the minimum value of the numerator in the square-root expression is $a-b$, (corresponding to $c=\min\{a,b\}$) while the maximum value is $a+b$ (corresponding to $c=0$). The range of values for Euc hence lies in the range

$$1 - \sqrt{\frac{a+b}{N}} \leq Euc \leq 1 - \sqrt{\frac{a-b}{N}}$$

It can be seen that if the value of a and b increases, then the left-hand expression decreases, which suggests that the starting point of the Euc values will decrease when the number of bits set becomes larger. Consider a dataset where a number of compounds have only a few bits set: then, if any compounds with a small number of bits set have already been selected in a DBCS procedure, it is unlikely that other compounds also having a small number of bits sets will be selected.

The Tanimoto coefficient behaves very differently, with the coefficient values generally increasing in line with an increase in the number of bits set in the molecules that are being compared. For the Tanimoto coefficient, Tan, the range of values is

$$0 \leq Tan \leq \frac{\min\{a,b\}}{\max\{a,b\}},$$

the minimum and maximum values corresponding to $c=0$ and $c=\min\{a,b\}$, respectively. Thus, in a similarity comparison, if one of the molecules is small, the upper bound of the Tanimoto, is small; large Tanimoto values are more likely to be produced if both bit-strings are densely populated (indicating large molecules). In the case of DBCS, therefore, smaller molecules have a greater possibility of being selected than do larger molecules. The Cosine coefficient exhibits the same behavior as the Tanimoto coefficient (see also (29)). The expression used for calculating the Cosine similarities, Cos, is

$$Cos = \frac{c}{\sqrt{ab}}$$

and by analogous arguments to those above, the range of values is

$$0 \leq Cos \leq \sqrt{\frac{\min\{a,b\}}{\max\{a,b\}}},$$

with larger molecules again tending to yield larger coefficient values. The upperbound values for *Tan* and *Cos* also suggest that the distribution of Cosine values will be broader than those for the Tanimoto coefficient. Now,

$$\frac{\min\{a,b\}}{\max\{a,b\}} \leq 1;$$

accordingly

$$\max\{Cos\} \geq \max(Tan)$$

with a consequent greater spread in the *Cos* values than in the *Tan* values as they both have the same lowerbound of zero. This is exactly what is observed in Figure 3.

APPLICABILITY OF THE SIMILAR PROPERTY PRINCIPLE

Thus far, we have considered the structural similarities between pairs of molecules, without considering any associated similarities in activity. Extensive experiments with fragment bit-strings have demonstrated that they support the similar property principle (*3-5*), but such experiments tend to focus on near neighbours; here, we look at the extent to which the similar property principle applies when the full range of similar values is considered. Specifically, consider a virtual screening environment where a known bioactive molecule is used as the target structure for a similarity search of a corporate database to prioritise molecules for biological testing. Then, we seek to determine the extent to which the principle applies as less and less similar molecules are included in the set of nearest neighbours for the target structure. If the similar property principle holds then the fraction of the database with the same activity as the target structure should decline steadily as the threshold similarity for inclusion in the nearest-neighbour is reduced.

Our experiments used the same two datasets and the same three fragment bit-string types as previously. For the AIDS dataset, which contains 247, active compounds, 802 moderately-active compounds and 4723 inactive compounds, the moderately-active compounds were classed first as inactive, and then as active, in a two-part study. For the ID Alert database, we selected ten bioactive types, for which the number of actives in each type ranged from 46 to 115, and used these as our active sets. In both cases we used each active molecule in turn as the target structure with the Tanimoto, Cosine, and Euclidean complement coefficients being used to rank the database in order of decreasing similarity with the target structure in each case. The percentage of the actives in each 0.05 percentile of the rankings was then noted.

The resulting plots (averaged over all of the target structures in each case and using UNITY 2D characterisations) are shown in Figure 4 for the AIDS database and Figure 5 for the ID Alert database. As expected, there is high percentage of actives in the 95-th percentile at the top of the rankings, but this is followed by a very rapid fall to the 90-th percentile in all the plots; there is then a much less marked variation in the percentage of actives (sometimes up and sometimes down but never by that much) till a final dip in the cases of the Cosine and Tanimoto coefficients. Plots such as these hence suggest that the similar property principle does indeed hold at the highest similarities (95-th percentile), and that its converse often holds at the lowest similarities (30-th to the 10-th percentile, depending on the database and bit-string characterization); at these very low similarities, therefore, a molecule that is chosen is indeed likely to have different biological characteristics from the target structure, supporting the use of such coefficients in DBCS-like applications. However, there is a large intermediate range of similarities where ranking by similarity will retrieve only a very few more, if any more, actives than would be obtained by random selection once the very similar molecules had been removed.

The Euclidean-complement plots in Figures 4c and 5c are very different as they appear, particularly in the case of the AIDS dataset, to be bimodal, with one peak at the 95-th percentile and one at around the 5-th to the 15-th percentile. This coefficient does not include a size normalization factor in the denominator (as do the Tanimoto and Cosine coefficients), and the presence of the $a+b-2c$ term thus results in an inherent bias towards larger molecules; specifically, the larger the molecules considered, the smaller the value of the Euclidean complement. Now both of the datasets have a fair number of active molecules that are large (i.e., that have many bits set) and these thus have an inherent *a priori* probability of having a low similarity with a target structure, with the resultant peaks in the lowest percentiles that are seen in Figures 4c and 5c.

The distributions shown in Figures 4 and 5 are pertinent to the statements that are sometimes made about the effectiveness or otherwise of different types of structure representation in similarity and diversity analyses. Many previous studies have shown that 2D fingerprints are very useful in such applications; however the distributions included here demonstrate that these favourable characteristics are observed only when certain similarity coefficients are employed. It is the combination of structure representation and similarity coefficient that together comprise a similarity measure that determines whether a successful analysis will be obtained.

CONCLUSIONS

Similarity measures based on chemical bit-strings and association or distance coefficients are widely used in similarity and diversity studies. In this paper, we have considered the magnitudes of the coefficient values that are obtained using three common similarity coefficients and three common bit-string representations. High coefficient values are known to encode meaningful relationships between different molecules and between structure and activity. Our results suggest that low coefficient values also encode significant relationships, although the utility of this relationship depends on the coefficient and on the size of the molecules that are used.

Acknowledgements.

We thank the following: Government of Malaysia for funding; Barnard Chemical Information Ltd., Daylight Chemical Information Systems Inc. and Tripos Inc. for software support; Current Drugs Ltd. for provision of the ID Alert dataset; and the Royal Society and Wolfson Foundation for hardware and laboratory facilities. The Krebs Institute for Biomolecular Research is a Biomolecular Sciences Centre of the Biotechnology and Biological Sciences Reseach Council.

REFERENCES

1. Willett, P.; Barnard, J. M.; Downs, G. M. *J. Chem. Inf. Comput. Sci.* **1998**, *38*, 983-996.
2. Dean, P. M.; Lewis, R.A. *Molecular Diversity in Drug Design*; Kluwer Academic Publishers: Dordrecht, 1999.

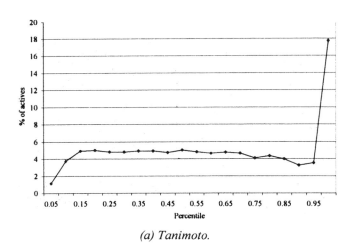

(a) Tanimoto.

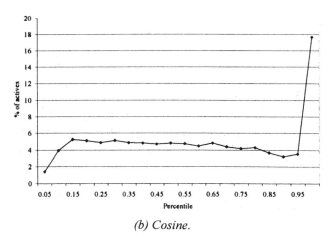

(b) Cosine.

Figure 4. Average % of actives covered in percentiles of similarity ranking in the AIDS database (UNITY 2D bit strings).

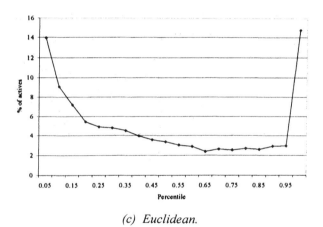

(c) Euclidean.

Figure 4. *Continued.*

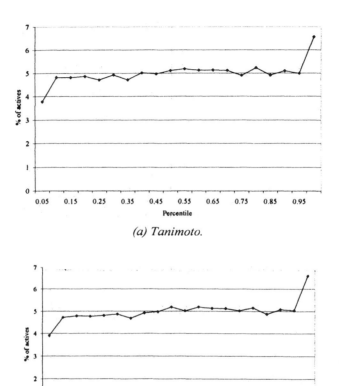

(a) Tanimoto.

(b) Cosine.

Figure 5. Average % of actives covered in percentiles of similarity ranking in the ID Alert database (UNITY 2D bit strings).

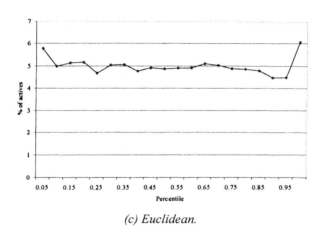

(c) Euclidean.

Figure 5. *Continued*

3. Brown, R. D.; Martin, Y. C. *J. Chem. Inf. Comput. Sci.* **1996**, *36*, 572-584.
4. Willett, P. *Similarity And Clustering In Chemical Information Systems*; Research Studies Press: Letchworth, 1987.
5. *Concepts and Applications of Molecular Similarity;* Johnson, M. A.; Maggiora, G. M., Eds.; Wiley: New York, 1990.
6. *Molecular Similarity in Drug Design*; Dean, P. M., Ed.; Chapman and Hall: Glasgow, 1994.
7. Adamson, G. W.; Bush, J. A. *J. Chem. Inf. Comput. Sci.* **1975**, *15*, 55-58.
8. Carhart, R. E.; Smith D. H.; Venkataraghavan, R. *J. Chem. Inf. Comput. Sci.* **1985**, *25*, 64-73.
9. Willett, P.; Winterman, V.; Bawden, D. *J. Chem. Inf. Comput. Sci.* **1986**, *26*, 36-41.
10. Lajiness, M. S. *Perspect. Drug Discov. Design.* **1997**, *7/8*, 65-84.
11. Flower, D. R. *J. Chem. Inf. Comput. Sci.* **1998**, *38*, 379-386.
12. Godden, J. W.; Xue, L.; Bajorath, J. *J. Chem. Inf. Comput. Sci.* **2000**, *40*, 163-166.
13. Dixon, S.L.; Koehler, R.T. *J. Med. Chem.* **1999,** *42*, 2887-2900.
14. Fligner, M.A.; Verducci, J.S.; Blower, P.E. *Technometrics* **2002**, *44*, 110-119.
15. *Combinatorial Library Design and Evaluation: Principles, Software Tools and Applications in Drug Discovery*; Ghose, A.K.; Viswanadhan, V.N., Eds.; Marcel Dekker: New York, 2001, pp. 379-398.
16. Mason, J.S.; Morize, I.; Menard, P.R.; Cheney, D.L.; Hulme, C.; Labaudiniere, R.F. *J. Med. Chem.* **1999**, *42*, 3251-3264.
17. Holliday, J.D.; Hu, C.Y.; Willett, P. *Combinatorial Chemistry & High Throughput Screening*, **2002**, *5*, 155-166.
18. Snarey, M.; Terret, N.K.; Willett, P; Wilton, D.J. *J. Mol. Graph. Model.* **1998**, *15*, 372-385.
19. Clark, R.D. *J. Chem. Inf. Comput. Sci.* **1997**, *37*, 1181-1188.
20. The NCI AIDS database is available at http://dtp.nci.nih.gov
21. The UNITY software system is produced by Tripos Inc. at http://www.tripos.com
22. Bradshaw, J.; Sayle, R. A. Available at URL http://www.daylight.com/meetings/emug97/Bradshaw/Significant_Similarity/Significant_Similarity.html
23. Sheridan, R.P.; Miller, M.D. *J. Chem. Inf. Comput. Sci.* **1998**, *38,* 915-924.
24. Adamson, G.W.; Lambourne, D.R.; Lynch, M.F. *J. Chem. Doc. C*, **1972**, 2428-2433.
25. Hodes, L.; *J. Chem. Inf. Comput. Sci.* **1976**, *16*, 88-93.

26. The ID Alert database is available from Current Drugs Ltd. at http://www.current-drugs.com
27. Barnard Chemical Information Ltd. is at http://www.bci1.demon.co.uk
28. Daylight Chemical Information Systems Inc. is at http://www.daylight.com
29. Holliday, J.D., Ranade, S. S. and Willett, P. *Quant. Struct.-Act. Relat.* **1995**, *14*, 501-506.

Chapter 7

Cheminformatics and Comparative Quantitative Structure–Activity Relationship

Rajni Garg

Chemistry Department, Clarkson University, Potsdam, NY 13699–5810

Cheminformatics encompasses the design, organization, storage, management, retrieval, analysis, dissemination, visualization and use of chemical information. Various tasks involved in cheminformatics are data mining, docking, defining quantitative structure-activity relationships (QSAR), pharmacophore mapping, and structure/substructure searching,. There are many approaches to QSAR and several programs available in the market. One such program is CQSAR that has been used to develop a CQSAR database that organizes QSAR from cheminformatics point in the area of physical-chemical and biological-chemical interactions. This database not only keeps account of what has been happening in the QSAR research but also provide numerous possibilities for comparing results from all kinds of biological systems (DNA, enzymes, receptors, cells, Human) with hydrophobic, electronic, and steric properties of the molecules. In this chapter the salient features of CQSAR program and database are considered in brief to illustrate its application in drug discovery, followed by a discussion of important physicochemical parameters (molecular descriptors) employed in developing QSAR. At the end "comparative QSAR of anti-HIV protease inhibitors" exemplify the role of cheminformatics and CQSAR in drug-design.

© 2005 American Chemical Society

Introduction

Discovery and marketing of a new drug costs a pharmaceutical company up to $650-800 million and takes an average of 12 to 15 years. Companies are continually exploring new software technologies and information resources to accelerate and streamline their drug-discovery activities. According to Brown "Cheminformatics is mixing of information resources to transform data into information, and information into knowledge, for the intended purpose of making better decisions faster in the arena of drug lead identification and optimization"(1). Various tasks involved in cheminformatics are data mining, docking, defining quantitative structure-activity relationships (QSAR), pharmacophore mapping, structure/substructure searching, and tools and approaches for predicting activity and other properties from structure.

Explosion of data from high throughput screening and combinatorial chemistry has put chemical information management and data analysis at the forefront of pharmaceutical research. Scientist involved in drug discovery process uses cheminformatics to define potential pharmaceutical agents or classes of agents, to be developed as potential lead. Synthesis of analogs of the lead molecule and their biological testing is an important aspect of drug design in order to obtain progressively better analogs. It is required not only for improving potency and efficacy but for pharmacokinetics (ADME) and side effects.

The principal hypothesis employed in the design of new analogs is that any change in the chemical structure produces a positive or negative change in the bioactivity. A systematic study of such cause and effect relation is called structure-activity relationship (SAR) study. SAR establishes the desirable changes in chemical structure and properties required for producing better biological activity. SAR has been made quantitative with the use of the physicochemical parameters (most often measured but sometimes calculated). The QSAR (quantitative structure-activity relationship) paradigm involves different quantitative approach to structure-property correlations in physical organic chemistry, biochemistry, and molecular design. It plays a vital role in lead exploitation (2,3).

Crum-Brown and Fraser (4) were the first to link physiological action to "chemical constitution". Meyer and Overton correlated biological activity with oil-water partition coefficients of narcotic substances (5,6). Albert established the importance of ionization and shape in the bacteriostatic action of aminoacridines (7). The most widely used and successful approach using descriptors is that of linear free energy relationship (LFER) as evidenced in the early work of Hammett (8) followed by excellent work of Hansch (2). Selassie et al. have described the development of QSAR in their recent work (9).

It is difficult to keep track of what is happening in the field of QSAR research. A search on QSAR results in more than 28,000 web sites. It is not possible to review all these pages and collect useful information. There are many approaches to QSAR, and several QSAR and SAR program available in the market. One such program is CQSAR (*10*), which has been used to generate a database of QSAR (CQSAR database) since its advent at Pomona College by Corwin Hansch and his group (*11*). CQSAR database organizes QSAR derived for the data obtained from the literature for physico-chemical (Phys-database) and bio-chemical (Bio-database) interactions. It is very important with respect to cheminformatics as it not only keeps account of what has been happening in the QSAR research but also facilitates the comparison of results from all kinds of biological systems (DNA, enzymes, cells, receptor, Human) with hydrophobic, electronic, and steric properties of the molecules.

Advances in understanding QSAR derived on biological system can be best understood by comparative studies (*12-16*). Comparative studies of a newly derived QSAR with other similar biological QSAR and the QSAR from mechanistic physical organic chemistry provide lateral support for its validation and helps in elucidating the reaction mechanism. In the following sections the salient features of the CQSAR program and database are considered in brief that illustrate its application in drug discovery, followed by a discussion of important physicochemical parameters employed in developing QSAR. At the end "comparative QSAR of anti-HIV protease drugs" exemplify the role of Cheminformatics and CQSAR in drug-design.

CQSAR Program and Database

The CQSAR program is used to develop and search CQSAR database. At present it contains over 18,500 equations that relate physico-chemical and bio-chemical properties of molecules to various molecular descriptors (physicochemical parameters). The data used to derive the QSAR are taken from various referred journals. Everyday new data are entered in the database. There are two databases in C-QSAR database:

- *Bio Database* – has more than 9,700 QSAR that correlate biological-chemical structure-activity relationship of biological systems,
- *Phys Database* – has more than 8,800 QSAR that correlate physical-chemical structure-activity relationship of physical organic systems.

Users can search the data in 20 different fields, e.g., by structure or substructure of the compounds studied, by the type of property correlated, by molecular properties, or by properties of the QSAR equation. The database is often used for data mining, to search for lead molecules, for substituent selection and "model mining" for lateral validation. The regression mode is used when the user wants to derive a new QSAR using new structures and activity data. For detail discussion on its use and applications, readers can refer to excellent articles by Hansch (17) and Kurup (18). A summary is given here.

Structure of database

- *Summary* - It is given with each dataset and provides information immediately about the data e.g., the type of parent molecule for which the specific activity (action) is studied, the reference, the number of compounds in the data set, the number of parameters used for regression analysis, number of compounds actually used in the QSAR etc. The output data describe the regression equation.
- *Classes and subclasses in Bio and Phys database* - The data is organized in such a way that one can request all the information pertaining to a particular problem for either a biological system or physical system. The biological data are categorized in 6 major classes and these are further divided into subclasses.
- *Physicochemical Parameters (molecular descriptors)* - The various molecular descriptors can be auto-loaded in the system for deriving QSAR (10). Broadly these parameters can be divided into three categories: hydrophobic, electronic and steric effects, which elucidate three important features of a chemical entity.

Searching the CQSAR database

Searching for the information that would aid drug design and development of new QSAR models that best 'explains' the preliminary activity data and provide lead for the synthesis of more active analogs are two most useful application of the database. QSAR from physical organic provides useful insight in understanding complex QSAR developed for biological system. Combined database search is useful feature of CQSAR in this direction. Three important mode of search are:

- *String search* – it is based on carefully selected words, accompanied by quote and blanks to narrow the search, e.g., a search on " BIOCHEM.J. " finds all the data from Journal of Biochemistry that is stored in the database.

- *Chemical structure/molecule search* - uses SMILES notation, developed for representing a chemical structure in 2D by Weininger (2,19,20).
- *Numeric search on the values of parameters* – compares various QSAR and aids in the selection of new substituents for the next round of synthesis that are expected to have better activity.

Substituent Selection in Molecular Design

One of the most significant and important feature of QSAR is to aid in substituents selection. A new QSAR guides in the selection of additional substituents, to design the next series of congeners, which will maximize the information obtained on enhanced performance. Parameter selection gives a list of substituents with known parameter values. Auto loading of any of these parameter values is unique to CQSAR.

A wise selection of substituents should involve sufficient variation so that there is enough spread in individual parameter values. Often it is noticed that sets reported in the literature are not usually designed keeping this factor in the mind. The QSAR that results from such data can be confusing or misleading (21).

Search for new lead in drug-design

- *QSAR searches on highly active compounds* - CQSAR database can be searched for the most active compounds exhibiting a specific activity. Most researchers made little attempt to formulate a QSAR using their structure-activity data. Almost all the bio QSAR in the database were developed by the Hansch group from published SAR data, therefore, it is unlikely that the most active congeners were discovered.
- *Substructure Search based on MERLIN search* - The MERLIN feature of the CQSAR database helps in searching for compounds by substructure searching (finger printing). Substructure search often finds too many examples for consideration. That can be made specific by entering a more specific name or SMILE.

Physicochemical Parameters (Molecular Descriptors)

There are numerous parameters used for describing hydrophobic, electronic and steric effects of a chemical entity. Although the CQSAR program can autoload more than 40 parameters, few are often found significant in delineating the important aspects of structure-activity profile of molecules. These parameters are:

- Hydrophobic – Clog P, π, MlogP
- Electronic – σ, σ^-, σ^+, σ_I, σ^*, F
- Steric – CMR, B1, L, B5, E_S, MgVol.

Hydrophobic Parameter (Clog P, π, MlogP)

Clog P is the calculated partition coefficient in octanol/water and is a measure of hydrophobicity of the whole molecule, while MlogP is the measured partition coefficient. There are various methods for calculating logP (22). The most extensively supported method is that of Leo (22,23). Following equation illustrate the quality of Leo's method: MlogP = 0.96 ($\pm$ 0.003) ClogP + 0.07 ($\pm$ 0.008); n = 12,510, r^2 = 0.973, q^2 = 0.973, s=0.300. The high correlation coefficient of r^2=0.973 assures one of the value of this method.

ClogP explains: (a) random walk process in movement of the drug molecule in the organism from site of injection to sites of action, (b) hydrophobic interactions between ligand and receptor. ClogP is for the <u>neutral form</u> of acids and bases that may be partially ionized. If the degree of ionization is about the same for a set of congeners one can neglect the ionization factor. If not, using electronic terms one can often obtain good correlations.

π is the hydrophobic parameter for substituents attached to benzene (2). Most of the published π constants are from the benzene system. For substituents with lone pair electrons (e.g. N), π for groups varies with electron effects of other substituents (e.g. π for OMe on benzene is not the same as OMe on pyridine or nitrobenzene). Calculating ClogP for a derivative and subtracting it from ClogP for the parent compound solve this problem. CQSAR program uses a parameter 'Cπ' that calculates the π value of substituents for systems other than benzene.

Electronic parameters (σ, σ^-, σ^+, σ_I, σ^*, F)

σ, σ^-, and σ^+ are Hammett electronic parameters, which apply to substituent effects on aromatic systems (8,24-28). The normal σ for substituents on aromatic systems where strong resonance between substituent and reaction center does not occur is defined as σ = logK_X–logK_H, where K_H is the ionization constant for benzoic acid (normally in water or in 50% ethanol) and K_X is that for substituted benzoic acid. σ^- and σ^+ are employed where there is a strong resonance interaction between substituent and reaction center (24,25). σ_I is a measure of inductive effect of aliphatic substituents (26). Taft σ^* applies electronic effects in aliphatic systems (27). F is the field (inductive) effect of an ortho substituent (28).

Steric Parameters (CMR, Bl, L, B5, E_S, MgVol)

CMR is the calculated molar refractivity for the whole molecule. MR is calculated by using Lorentz-Lorenz equation as follows: $(n^2 - 1/n^2 + 2)$ (MW/d), where n is the refractive index, MW is the molecular weight, and d is the density of a substance. MR is dependent on volume and polarizability. In CQSAR, MR values have been scaled by 0.1. MR can be used for a substituent or for the whole molecule (2). Recently, CMR along with NVE (net valence electrons) have also been found to account for the polarizability effects in ligand-substrate interactions (29).

B1, B5, and L are the Verloop's sterimol parameters for substituents (30). B1 is a measure of the width of the first atom of a substituent, B5 is an attempt to define the overall volume, and L is the substituent length.

E_S is the Taft steric constant (31). It is based on the acid catalyzed hydrolysis of α-substituted acetates, and represents the steric effect of intramolecular and intermolecular bulk, which hinders the reaction or binding. MgVol is the molar volume calculated by the method of McGown (32).

Comparative QSAR Studies on Anti-HIV-1 Protease Drugs

AIDS is the end stage manifestation of prolonged infection with HIV, particularly HIV-1, the most common form of the virus. Medicinal chemists have explored all the intervention stages in the viral life cycle to develop anti-HIV-1 chemotherapy. HIV-1-protease inhibitors, in particular, has drastically decreased the mortality and morbidity associated with AIDS. The determination of three-dimensional structure of HIV-1 protease, obtained through X-ray crystallography and the synthesis of the protease facilitated the development of HIV-1 protease inhibitors (33,34,35,36).

At present CQSAR database contains 225 QSAR on HIV-1 inhibitors. Out of which 70 are on HIV-1 protease. Cyclic urea (Figure 1) based non-peptidic HIV-1 protease inhibitors have been very well studied from SAR and QSAR point. Comparative study of cyclic urea QSAR models drawn from CQSAR database and literature discussed here provide an excellent example of integration of QSAR and Cheminformatics (37).

QSAR 1 and 2 was formulated for the data of Lam et al. (*38*) reported for the inhibition of HIV-1 protease (K_i) and antiviral activity (IC_{90}) of cyclic urea derivatives (Figure 2), in which X-substituents were mostly alkyl groups.

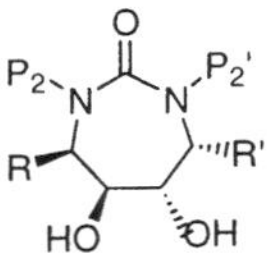

Figure 1

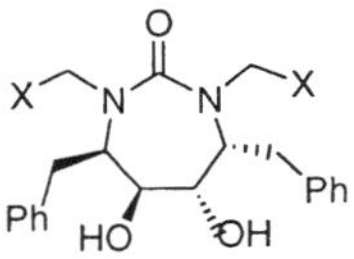

Figure 2

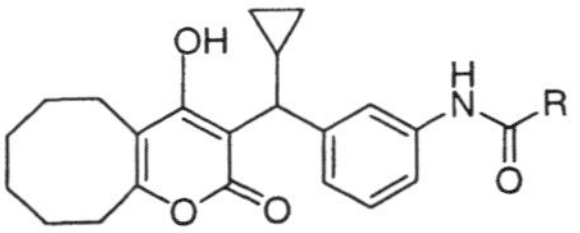

Figure 3

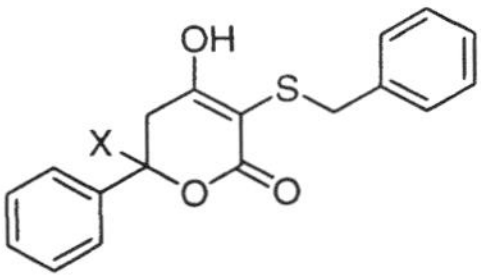

Figure 4

$$\log 1/K_i = 1.44(\pm 0.42)\text{ClogP} - 2.13(\pm 0.64)\log(\beta.10^{\text{ClogP}} + 1) + 0.68(\pm 0.42)\text{MR}_X - 0.64(\pm 2.22)$$

$n = 21, r^2 = 0.813, s = 0.51, q^2 = 0.727, \log P_0 = 6.53$, Outliers: 5 (1)

$$\log 1/IC_{90} = 0.77(\pm 0.25)\text{ClogP} - 1.24(\pm 0.48)\log(\beta.10^{\text{ClogP}} + 1) + 1.05(\pm 1.37)$$

$n = 15, r^2 = 0.813, s = 0.33, q^2 = 0.665, \log P_0 = 6.96$, Outliers: 2 (2)

In QSAR 1 and 2, K_i is the HIV-1 protease inhibition constant and IC_{90} is the molar concentration of the compound required to reduce the concentration of HIV viral RNA by 90% from the level measured in an infected culture. n, is the number of data points in the particular dataset studied, r is the correlation coefficient, s is the standard deviation, q is the quality of fit, and the data within parentheses are for the 95% confidence intervals. LogP_0 is the optimum value of logP. Outliers indicate the number of the molecules of the data set that are not fit in the QSAR model. The presence of outliers in QSAR is discussed later.

Both QSAR models show bilinear dependence of the inhibitory activity on the ClogP. This means that the inhibitory activity decreases if a molecule is more hydrophobic than its optimum value ($\log P_0 = 6.53$ and 6.96 respectively for both the series). In the first QSAR, MR_X was also found significant for the X-substituents at P_2 and P_2' position of cyclic urea. MR is the molar refractivity of the substituent and is a measure of its size and polarizability.

Out of 14 QSAR models reported for different class of cyclic urea HIV-1 protease inhibitors (37), only the two shown here contain hydrophobic terms. Interestingly all the other cyclic urea in which there was no hydrophobic term, contain X-benzyl substituents at P/P' position (Figure 1). Maybe the rigid phenyl ring blocks the substituents from interacting with the hydrophobic space of the active site. However, a significant steric interaction with the receptor seems to be involved in almost all of them as evident by the occurrence of steric terms in other QSAR (37).

Comparative molecular field analysis (CoMFA) of diverse cyclic urea HIV-1 protease inhibitors also established steric and electrostatic interactions between ligand and receptor binding site (39). The same studies revealed significant contribution of hydrophobicity toward protease inhibitory activity.

Out of 12 QSAR reported on pyranones (37), only two-showed hydrophobic term (QSAR 3 and 4). QSAR 3 was based on Romines et al. (40) data for cycloalkylpyranones derivatives (Figure 3), and QSAR 4 was based on Tait et al. (41) data for dihydropyranones (Figure 4) for HIV-1 protease inhibition.

$$\log 1/K_i = 5.44(\pm 3.81)\text{ClogP} - 0.61(\pm 0.37)(\text{ClogP})^2 - 3.84(\pm 9.58)$$

$n=9$, $r^2=0.910$, $s=0.24$, $q^2=0.770$, $\log P_0 = 4.49$, Outliers: 2 \hfill (3)

$$\log 1/\text{IC}_{50} = 0.82(\pm 0.32)\text{ClogP} - 0.44(\pm 0.20)B5_X + 4.05(\pm 1.11)$$

$n=10$, $r^2=0.841$, $s=0.12$, $q^2=0.742$, Outlier: 1 \hfill (4)

In QSAR 4, the $B5_X$ term was also found significant. B5 is Verloop's width parameter (30) and indicates the steric interactions of the substituents at the active site.

Except for these two QSAR shown here, others did not reveal the presence of any hydrophobic term (37). Garg et al. postulated that may be due to some spatial restrictions these molecules are not able to bind in hydrophobic space as the protease receptor does have hydrophobic binding sites. The optimum size of the cycloalkyl ring should be 8-membered indicate maximal hydrophobic interaction of the ring with the receptor (42).

HIV-protease is a C_2 symmetrical homodimer (33). The C_2 axis of the enzyme lies between and perpendicular to catalytic aspartates (Asp 25 and Asp 25') in the active site. The S_1 and S_1' (S_2 and S_2' etc) subsites are structurally identical and hydrophobic (*43,44*). It has also been proposed that there are multiple hydrogen bonding possibilities with HIV-1 protease receptors (*44*). Hydrophobic as well as hydrogen bond interactions have been shown to be equally significant in HIV-1 protease inhibition by QSAR studies (*45*). Wang et al. (*46*) observed that at least two factors are important in the binding of a compound to HIV-1 protease. The first is the conformational flexibility of the inhibitor molecule and the second is the hydrophobic interactions between an inhibitor and the enzyme. Favorable interactions with hydrophobic pockets at active site are desirable for an inhibitor to achieve nanomolar potency (*47*).

The ClogP values of U.S. Food and Drug Administration (FDA) approved HIV protease drugs in the market and optimum logP observed in the QSAR models were found surprisingly close. The stereo diagram of binding of a cyclic urea protease inhibitor in the HIV-1 protease pocket indicated that the phenyl groups in the cyclic urea inhibitors make hydrophobic interactions with the hydrophobic residues (*37*).

FDA Approved Protease Inhibitors	ClogP*
1. Saquinavir (Invirase®)	4.73
2. Ritanovir (Norvir®)	4.94
3. Indinavir (Crixivan®)	3.68
4. Nelfinavir (Viracept®)	5.84
5. Amprenavir (Agenerase®)	3.29

*Calculated using CQSAR program (10)

QSAR equation	Optimum LogP ($logP_0$)
1	6.53
2	6.96
3	4.49

The majority of HIV research is done with cells and these studies tend to over estimate $logP_0$ for animal systems. From a study of CQSAR database (omitting QSAR based on charged molecules), it has been found that $logP_0$ for cells is about 2 log unit higher than for whole organisms (*10*). Thus values above 4 for $logP_0$ may be too high.

Most QSAR models developed for HIV-1 protease inhibitors did not show hydrophobic terms, when there are certainly hydrophobic binding sites at the receptor. It could be due to the fact that most of the equations were based on a small number of data points. The variation in the substituents also does not allow for much choice in the use of different physicochemical parameters. A minimum of five data points with good variation in substituents is required per parameter to derive a meaningful QSAR.

QSAR analysis often shows presence of the outliers ('congeners' that do not fit the 'final' QSAR). Some of the reasons could be:

1. The mathematical form of the equation may not reflect the true interactions at the binding site.
2. Sometimes the parameters selected may not be the best. Lateral validation using comparative QSAR guides in the choice of meaningful parameters (*13-15,37,49*).
3. The quality of the experimental data coming from different laboratory affects the quality of QSAR. This problem can be taken care by comparing several QSAR models derived for the same systems (*12*).
4. The outliers that seem to be 'congeners' but in fact are not. This arises from trying to lump too many more or less similar compounds into a single QSAR (*16,50*).
5. Different rates of metabolism of the members of a set affect the activity data. The uncertainty that exists in describing what all is occurring with testing of a set of 30-40 'congeners' in even a simple cell culture is enormous (*14*).

Therefore, one has to expect outliers that must not be forgotten for they are the leads to new understanding. To cover them up by including them in a QSAR can be more confusing than helpful.

Prolonged use of HIV protease drugs have lead to the development of resistant mutant strains that are less sensitive to the inhibitors (*48*). The emergence of mutant virus and occurrence of side effects suggest that it is essential to continue the research to develop new inhibitors.

A number of structure-activity analyses using both 2D classical QSAR (*37,45,49,51-53*) and 3D QSAR approaches (*39,54-57*) have provided insight in the ligand-HIV-1-protease interactions involving different classes of inhibitors. Due to the limitations of every single structure-activity approach, a combination of 2D and 3D QSAR approaches is important to understand complex phenomenon contributing to anti-HIV-1 protease activity. Lateral validation of a biological QSAR from a physical organic viewpoint and added biological perspective, makes CQSAR an important approach in cheminformatics for mechanistic interpretation, and can provide important lead(s) in drug development.

Acknowledgement

I thank Corwin Hansch and Alka Kurup for many helpful suggestions and critical reading of the manuscript.

References

1. Brown, F. K. *Annu. Rep. Med. Chem.* 1998, *33*, 375-384.
2. Hansch, C.; Leo, A. *Exploring QSAR. Fundamentals and Applications in Chemistry and Biology*; American Chemical Society: Washington, DC, 1995.
3. Hansch, C.; Leo, A.; Hoekman, D. *Exploring QSAR: Hydrophobic, Electronic, and Steric Constants*; American Chemical Society: Washington, DC, 1995.
4. Rekker, R.F. *Quant. Struct.-Act. Relat.* 1992, *11*, 195-199.
5. Meyer, H. *Arch. Exp. Pathol. Pharmakol.* 1899, *42*, 109-118.
6. Overton, E. *Z. Physik. Chemie.* 1897, *22*, 189-209.
7. Albert, A.; Rubbo, S.D.; Goldacre, R. *Nature* (London) 1941, *147*, 332-333.
8. Hammett, L.P. *Physical Organic Chemistry*; McGraw-Hill: New York, 1940.
9. Selassie, C.D.; Mekapati, S.B.; Verma, R.P. *Current topics Med. Chem.*, 2002, *2*, 1357-1379.
10. CQSAR program, BioByte Corp., 201 W, 4th St. Suite 204, Claremont, CA 91711. (www.biobyte.com)
11. Hansch, C.; Maloney, P.P.; Fujita, T.; Muir, R.M. *Nature* (London) 1962, *194*, 178-180.
12. Selassie, C.D.; Garg, R.; Kapur, S.; Kurup, A.; Verma, R.P.; Mekapati, S.B.; Hansch, C. *Chem. Rev.* 2002, *102*, 2585-2605.
13. Kurup, A.; Garg, R.; Hansch, C. *Chem. Rev.* 2001, *101*, 2573-2600.
14. Hansch, C.; Kurup, A.; Garg, R.; Gao, H. *Chem. Rev.* 2001, *101*, 619-672.
15. Gao, H.; Katzenellenbogen, J.A.; Garg. R.; Hansch, C. *Chem. Rev.* 1999, *99*, 723-45.
16. Garg, R., Kapur, S., Hansch, C. *Med. Res. Rev.* 2000, *21*, 73-82.
17. Hansch, C.; Hoekman, D.; Leo, A.; Weininger, D.; Selassie, C.D. *Chem. Rev.* 2002, *102*, 783-812.
18. Kurup, A. *J. Comp. Aided Mol. Des.* 2003 (in press)
19. Weininger, D.; Weininger, A.; Weininger, J.L. *J. Chem. Inf. Comp. Sci.* 1989, *29*, 97-101.

20. Weininger, D.; Weininger, J.L. In *Comprehensive Medicinal Chemistry*; Hansch, C.; Sammes, P.G.; Taylor, J.B., Ed.; Pergamon Press: Oxford, NewYork; 1990; Vol. 4., pp. 59-82.
21. Mekapti, S.B.; Hansch, C. *J. Chem. Inf. Comp. Sci.*,2002, *42*, 956-961.
22. Leo, A. *Chem. Rev.* 1993, *93*, 1281-1306.
23. Leo, A. and Hansch, C. *Perspect.* Drug Discov. Des. 1999, *17*, 1-25.
24. Hammett L.P. Chem. Rev. 1935, 17, 125-136.
25. Brown, H.C.; Okamto, Y. *J. Am. Chem. Soc.* 1958, *80*, 4979-4987.
26. Ingold,C.K. *Structure and Mechanism in Organic Chemistry 2nd Ed*; Cornell University Press: Ithaca, NY, 1969.
27. Taft, R.W. Jr. *J.Am.Chem.Soc.* 1958, *80*, 2436-2443.
28. Swain, C.G.; Lupton, E.C. Jr. *J. Am. Chem. Soc.* 1968, *90*, 4328-4337.
29. Hansch, C; Steinmetz,W.E.; Leo, A.J.; Mekapati, S.B.; Kurup, A.; Hoekman, D. *J. Chem. Inf. Comput. Sci.* 2003, *43*, 120-125.
30. (a) Verloop, A. *The Sterimol approach to Drug Design*; Marcel Dekker: New York, 1987. (b) Verloop, A.; Hoogenstraaten, W.; Tipker, J. In *Drug Design*: Ariens, E.J., Ed.; Academic Press: New York, 1976; Vol VII, pp. 165-207.
31. Taft, R.W. *Steric effects in organic Chemistry*: Newman, M.S., Ed.; Wiley: New York, 1956.
32. Abraham, M.; McGown, J.A. *J. Chrmotatographica*, 1987, 23, 243-244.
33. Toh, H.; Ono, M.; Saigo, K.; Miyata, T. *Nature* (London) 1985, *316*, 21-22.
34. Navia, M.A.; Fitzgerald, P.M.; McKeever, B.M.; Leu, C.T.; Heimbach, J.C.; Herber, W.K. Sigal. I.S.; Darke, P.L.; Springer, J.P. *Nature* (London) 1989, *337,* 615-620.
35. Wlodawer, A.; Miller, M.; Jaskolski, M.; Sathyanarayana, B.K.; Baldwin, E.; Weber, I.T.; Selk, L.M.; Clawson, L.; Schneider, J.; Kent, S.B. *Science* 1989, *245*, 616-621.
36. Flexner C. *N. Engl. J. Med.* 1998, *338,* 1281-1292.
37. Garg, R.; Gupta, S.P.; Gao, H.; Mekapati, S. B.; Debnath, A.K.; Hansch, C. *Chem. Rev.* 1999, *99*, 3525-3602.
38. Lam, P.Y.; Ru, Y.; Jadhav, P.K.; Aldrich, P.E.; DeLucca, G.V.; Eyermann, C.J.; Chang, C.H.; Emmett, G.; Holler, E.R.; Daneker, W.F.; Li, L.; Confalone, P.N.; McHugh, R.J.; Han, Q.; Li, R.; Markwalder, J.A.; Seitz, S.P.; Sharpe, T.R.; Bacheler, L.T.; Rayner, M.M.; Klabe, R.M.; Shum, L.; Winslow, D.L.; Kornhauser, D.M.; Hodge, C.N. et al. *J. Med. Chem.* 1996, *39*, 3514-3525.
39. Debnath, A.K. *J. Med. Chem.* 1999, *42,* 249-259.
40. Romines, K.R.; Watenpaugh, K.D.; Howe, W.J.; Tomich, P.K.; Lovasz, K.D.; Morris, J.K.; Janakiraman, M.N.; Lynn, J.C.; Horng, M.M.; Chong, K.T. et al. *J. Med. Chem.* 1995, *38*, 4463-4473.

41. Tait, B.D.; Hagen, S.; Domagala, J.; Ellsworth, E.L.; Gajda, C.; Hamilton, H.W.; Prasad, J.V.; Ferguson, D.; Graham, N.; Hupe, D.; Nouhan, C.; Tummino, P.J.; Humblet, C.; Lunney, E.A.; Pavlovsky, A.; Rubin, J.; Gracheck, S.J.; Baldwin, E.T.; Bhat, T.N.; Erickson, J.W.; Gulnik, S.V.; Liu, B. *J. Med. Chem.* 1997, *40*, 3781-3792.
42. Skulnick, H.I.; Johnson, P.D.; Aristoff, P.A.; Morris, J.K.; Lovasz, K.D.; Howe, W.J.; Watenpaugh, K.D.; Janakiraman, M.N.; Anderson, D.J.; Reischer, R.J.; Schwartz, T.M.; Banitt, L.S.; Tomich, P.K.; Lynn, J.C.; Horng, M.M.; Chong, K.T.; Hinshaw, R.R.; Dolak, L.A.; Seest, E.P.; Schwende, F.J.; Rush, B.D.; Howard, G.M.; Toth, L.N.; Wilkinson, K.R.; Romines, K.R. et al. *J. Med. Chem.* 1997, *40*, 1149-1164.
43. Schechter, I.; Berger, A. *Biochem. Biophys. Res. Commun.* 1967, *27*, 157-162.
44. Babine, R.E.; Bender, S.L. *Chem. Rev.* 1997, *97*, 1359-1472.
45. Gupta, S.P.; Mekapati, S.B.; Garg, R.; Sowmya, S. *J. Enzyme. Inh.* 1998, *13*, 399-407.
46. Wang, S.; Milne, G.W.; Yan, X.; Posey, I.J.; Nicklaus, M.C.; Graham, L.; Rice, W.G. *J. Med. Chem.* 1996, *39*, 2047-2054.
47. Bernstein, F.C.; Koetzle, T.F.; Williams, G.J.; Meyer, E.F. Jr.; Brice, M.D.; Rodgers, J.R.; Kennard, O.; Shimanouchi, T.; Tasumi, M. *Arch. Biochem. Biophys.* 1978, *185*, 584-591.
48. Ridky,T.; Leis, J. *J. Biol. Chem.* 1995, *270*, 29621-29623.
49. Kurup, A.; Mekapati, S.B.; Garg, R.; Hansch C. *Current Topics Med. Chem.* 2003 (in press).
50. Selassie, C.D.; DeSoyza, T.V.; Rosario, M.; Gao, H.; Hansch, C. *Chem. Biol. Interact.* 1998, *113*, 175-190.
51. Wilkerson, W.W.; Akamike, E.; Cheatham, W.W.; Hollis, A.Y.; Collins, R.D.; DeLucca, I.; Lam, P.Y.; Ru, Y. *J. Med. Chem.* 1996, *39*, 4299-4312.
52. Wilkerson, W.W.; Dax, S.; Cheatham, W.W. *J. Med. Chem.* 1997, *40*, 4079-4088.
53. Wei, D.T.; Meadows, J.C.; Kellogg, G.E. *Med. Chem. Res.* 1997, *7*, 259-270.
54. Waller, C.L.; Oprea, T.I.; Giolitti, A.; Marshall, G.R. *J. Med. Chem.* 1993, *36*, 4152-4160.
55. Opera, T.I.; Waller, C.L.; Marshall, G.R. *Drug. Des. Discov.* 1994, 12, 29-51.
56. Oprea, T.I.; Waller, C.L.; Marshall, G.R. *J. Med. Chem.* 1994, *37*, 2206-2215.
57. Debnath, A.K. J. Chem. Inf. Comput. Sci. 1998, 38, 761-767.

Chapter 8

Prediction of Protein Retention Times in Anion-Exchange Chromatography Systems Using Support Vector Regression

Curt M. Breneman[1,*], Minghu Song[1], Jinbo Bi[2], N. Sukumar[1], Kristin P. Bennett[2], Steven Cramer[3], and N. Tugcu[3]

Departments of [1]Chemistry, [2]Mathematics, and [3]Chemical Engineering, Rensselaer Polytechnic Institute, 110 8th Street, Troy, NY 12180
[*]Corresponding author: telephone: 518–276–2678; fax: 518–276–4887; email: brenec@rpi.edu

ABSTRACT

Quantitative Structure-Retention Relationship (QSRR) models are developed for predicting protein retention times in anion exchange chromatography. Constitutional, topological and electron-density-based descriptors are computed directly from the protein crystal structures. These QSRR models are constructed based on the Support Vector Regression (SVR) algorithms. To accomplish this, a two-step computational strategy was adopted. In the first step, a linear SVR was utilized as a variable selection method and the relative importance of selected descriptors is analyzed using the star plot visualization approach. Subsequently, the selected features are used to produce nonlinear SVM bagged models. After validation, these predictive models may be used as part of an automated virtual high-throughput screening (VHTS) process.

© 2005 American Chemical Society

I. INTRODUCTION

Ion-Exchange Chromatography (IEC) is one of important bioseparation technique at the heart of drug discovery and developing biotechnology. This form of chromatography takes advantage of the fact that the biopolymers, such as proteins, have different degree of interaction with the oppositely charged resin surface; this lead to their different retention characteristics during the separation process. The selectivity of this technique is also depended on the experimental environment, such as composition of the stationary phase or the pH of the mobile phase. Consequently, one of the challenges is to select appropriate combinatorial conditions as so to achieve best purification for a given biological mixture.

It has been suggested that virtual screening (VS) of separation materials or experimental condition in a manner that parallels current lead discovery techniques in drug design would probably facilitate bioseparation development processes. Different statistical algorithms, such as Principal Component Regression (PCR)[1], Partial Least Squares (PLS)[2,3], and Artificial Neural Networks (ANN)[4,5], have been employed to construct the Quantitative Structure-Retention Relationship (QSRR) models[6] within the chromatography community. In the current study, a novel modeling approach based on Support Vector Machine (SVM) Regression[7] was present to predict the retention time of proteins in anion exchange systems. Meanwhile, a visualization tool, the star plot, is employed to aid in model interpretation. The test data that hold out during the training process is used to validate the predictive power of the constructed models. The objectives of our study are to construct improved QSRR models to predict the retention behavior of proteins in specific experimental conditions and choose the optimal separation condition to achieve best purification, as well as to build a valuable interpretation tool for the protein retention mechanisms.

II. SUPPORT VECTOR REGRESSION (SVR)

SVMs[8] are a class of supervised learning algorithms originally developed for pattern recognition and later its basic properties were extended by Vapnik to solve the regression problems. Given the target value y_i and the hypothesis space consisting of linear functions in the form of $\langle w \cdot x \rangle + b$, the classical SVR try to find a function $f(x_i)$ that minimizes the overall regularized risk [9]:

$$C \sum_{i=1}^{M} |y_i - f(x_i)|_\varepsilon + \frac{1}{2} \|w\|^2 \qquad (1)$$

In the above formula, the first term $\sum_{i=1}^{M} |y_i - f(x_i)|_\varepsilon$ computes the training error and the second term, the l_2-norm $\frac{1}{2}\|w\|^2$ of normal vector, controls the model complexity. The C is a regularization parameter determining the tradeoff between training error and model complexity. In support vector machine, the training error is represented by the ε-insensitive losses (illustrated in figure 1), in which only those deviations larger than the tolerance errors would be considered as errors. The magnitude of ε would be roughly the estimated error in experimental measurement. In the regularization factor $\frac{1}{2}\|w\|^2$, ω is the weight vector to be determined in the function f. The SVR problem can be posed as a convex optimization problem as follows:

$$\text{minimize} \quad C\sum_{i=1}^{M}(\xi_i+\xi_i^*)+\frac{1}{2}\|w\|^2 \qquad (2)$$

$$\text{subject to} \quad y_i-\langle w\cdot x_i\rangle-b\leq \varepsilon+\xi_i, \xi_i\geq 0$$
$$\langle w\cdot x_i\rangle+b-y_i\leq \varepsilon+\xi_i^*, \xi^*\geq 0$$
$$i=1,2,\ldots\ldots,M$$

ξ and ξ* stand for the slack variables that measure the deviation distance from the data point to the ε tube.

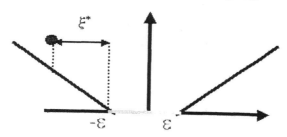

Figure 1. *Graphical depiction of an ε-insensitive loss function. Only the deviations of data points larger than ε, such as ξ*, will be considered as the errors.*

Another main characteristics of SVR is to map the original data point x into a higher dimensional feature space via the nonlinear kernel function $k(x_i, x)$, and then perform the linear regression in that high dimensional feature space so as to achieve the nonlinear relationship with respect to the original input space. Now the function f can be written in the form of a kernel expansion as:

$$f(x) = \sum_{i=1}^{M} \alpha_i k(x_i, x) + b \tag{3}$$

III. DATASET GENERATION

Protein Retention Dataset

There are two main kinds of IEC system: cation-exchange and anion-exchange chromatography. This study will only focus on the anion-exchange system, which bears the positively charged functional groups on its resin surface. The crystal structures of 24 structurally diverse proteins were downloaded from the RSCB Protein Data Bank[10]. The retention times for these proteins were obtained by carrying out linear gradient chromatography using the anion exchange stationary phase Source 15Q. Three proteins were randomly selected as external test cases from this original list.

Descriptor Generation

As we known, the direct calculation of quantum chemical descriptor for large molecules at a high level of theory is usually prohibitive. The concept of "Transferable Atom Equivalents" (TAEs)[11,12], based on Bader's quantum theory of Atoms in Molecules (AIM)[13], provides an efficient alternative to calculate these electron-density derived descriptors. Transferable Atom Equivalents are defined as those atomic electron density fragments bounded by interatomic zero-flux surfaces ($\nabla \rho(r) \cdot n(r) = 0$, for all points on the surface) and an extended isodensity surface that approximates the condensed-phase van der Waals surface. An atomic property (A) can then be expressed as the integral of a corresponding property density $\rho_A(r)$ over an atomic basin:

$$A(\Omega) = \int_\Omega d\tau \rho_A(r) \text{ where } \rho_A(r) = (N/2) \int d\tau' \left\{ \psi^* \hat{A} \psi + \left(\hat{A} \psi \right)^* \psi \right\} \tag{4}$$

TAE fragments carry ten atomic charge density-derived properties (listed in Table 1) that were pre-computed from small molecules using *ab initio* wave functions at the 6-31+G* level of theory. The distributions of these electronic properties computed on electronic density isosurfaces may be characterized as molecular property descriptors by histogram binning, averaging or sampling the property extrema or standard deviation. The RECON (RECONstruction) program assigns the closest fragment match from the TAE library to each atom in the protein based on its structural and chemical environment. Consequently, a large set of electron density-based TAE descriptors for proteins can be obtained by summing up the corresponding electronic properties of transferable atomic fragments: $A_{molecule} = \sum_\Omega A(\Omega)$ These descriptors contain certain information about molecular basicity, hydrophobicity, hydrogen-bonding capacity and polarity as well as polarizability.

The MOE program provides a widely applicable set of classical molecular descriptors, including traditional physicochemical properties, connectivity-based topological 2D and shape-dependent 3D molecular features. These descriptors have been applied to the construction of QSAR/QSPR models for boiling point,

Table 1. TAF Atomic Electronic Surface Properties

EP	Electrostatic Potential
Del(Rho) • N	Electron Density Gradient normal to 0.002 e/au^3 electron density isosurface
G	Electronic Kinetic Energy Density $G = \left(-(\eta/4m)\int\{\nabla\psi^* \cdot \nabla\psi\}d\tau\right)$
K	Electronic Kinetic Energy Density $K = -(\eta/4m)\int\{\psi^*\nabla^2\psi + \psi\nabla^2\psi^*\}d\tau$
Del(K) • N	Gradient of K Electronic Kinetic Energy Density normal to Surface
Del(G) • N	Gradient of B Electronic Kinetic Energy Density normal to surface
Fuk	Fukui F$^+$ function scalar value
Lapl	Laplacian of the electron density $\nabla^2\rho$
BNP	Bare Nuclear Potential $BNP_{(j)} = \sum_{i=1}^{n} q_i / r_{ij}$
PIP	Local Average Ionization Potential $PIP(r) = \sum_i \rho_i(r)|\varepsilon_i|/\rho(r)$

vapor pressure, the free energy of solvation in water, as well as water solubility and blood-brain barrier penetration.[14]

A total of 243 descriptors, including electron density-derived and traditional descriptors, were then computed for these proteins and subjected to SVR training and cross-validation experiment.

IV. METHOD IMPLEMENTATION

l_1-norm v-SVR and Implementation Strategy

In this study, we adopt a variation of the classical SVR called v support vector regression (v - SVR).[15,16] In v-SVR, ε itself is a variable in the optimization process and is controlled by another new parameter $v \in (0,1)$. The v is the upper bound on the fraction of error points and the lower bound on the fraction of points inside the ε-insensitive tube, which is much easier to estimate beforehand than ε. In addition, in order to reduce the computational cost, a linear program, instead of a quadratic program, is formulated for SVR. The l_1-norm is applied directly to the coefficients $\alpha_j, j = 1,...M$ in the kernel combination Eq. (3). It can be computed as $\sum_{j=1}^{M} |\alpha_j|$ or rewritten as $\sum_{j=1}^{M} (\alpha_j + \alpha_j^*)$ if we define $\alpha_j = \alpha_j - \alpha_j^*$, where $\alpha_j \geq 0$ and $\alpha_j^* \geq 0$. Substitution of $\frac{1}{2}\|w\|^2$ by $\sum_{j=1}^{M} (\alpha_j + \alpha_j^*)$ yields the linear program.

A two-step computational strategy was adopted: First, a l_1-norm linear SVR was utilized as a variable selection method to identify relevant molecular descriptors: second, a set of nonlinear SVR models derived based on kernel mapping were constructed using the selected features. In addition, a statistical technique called "bagging" (Bootstrap Aggregation) was applied to improve model generalization performance.

Feature Selection

For most QSAR or QSRR data, one of common problems is that the number of observations is much fewer than that of descriptors. So it is essential to utilize efficient feature selection or regularization methods to remove irrelevant descriptors and increase the accuracy. Also it will speed up the whole learning process and make interpretation easier by emphasizing only a few relevant features. A variety of algorithms, such as forward selection,[17] simulated annealing,[18] genetic algorithms,[19,20] K-nearest neighbor,[21] evolutionary programming,[22,23] artificial ants,[24,25] and binary particle swarms[26], have been implemented in the QSAR or QSPR studies.

The feature selection method used in this work exploits the fact that linear SVR with l_1-norm regularization inherently performs feature selection as a side effect of minimizing function capacity during the modeling process.[27] In a linear regression model in the form of $y = \langle \alpha \cdot x \rangle + b$, each component of vector α provides a weight for the corresponding feature, thus providing a measure of its significance in the model. Moreover, the sign of each component α_i indicates the effect of the ith feature on the observed response. If $\alpha_i > 0$, the feature contributes positively to the response y, and when negative it diminishes y. Training support vector machines involves maximization of the "margin", a term that is inversely proportional to the norm of the weights $\|\alpha\|$. The margin is defined as the geometric width of the ε-tube and it provides a measure of model complexity. Maximizing the margin (or minimizing the norm of the weights) implicitly makes the optimal weight vector sparser. A vector is sparse if the number of non-zero components (descriptor weights) in the vector is small. This method of feature selection is formulated as an l_1-norm linear ν-SVR aimed at producing a sparse weight vector. We refer to it as the sparse linear ν-SVR, and it can be stated in the following manner:

$$\text{minimize} \quad \frac{1}{2}\sum_{j=1}^{N}(\alpha_j + \alpha_j^*) + C\frac{1}{M}\sum_{j=1}^{M}(\xi_i + \xi_i^*) + C\nu\varepsilon \quad (5)$$

$$\text{subject to} \quad y_i - \sum_{j=1}^{N}(\alpha_j - \alpha_j^*)x_{ij} + b - y_i \leq \varepsilon + \xi_i i, \quad i = 1, 2, \ldots, M$$

$$\sum_{j=1}^{N}(\alpha_j - \alpha_j^*)x_{ij} + b - y_i \leq \varepsilon + \xi_i,$$

$$\alpha_j, \alpha_j^*, \xi_i, \xi_i, \varepsilon \geq 0, \quad j = 1, 2, \ldots, N, \quad i = 1, 2, \ldots, M$$

The l_1-norm SVR optimization can enhance the sparsity of α because it more intends to drive the weights of irrelevant descriptors to zero. Those descriptors with nonzero weights then become potentially relevant features to be selected and used to build a subsequent nonlinear model.

In the case of much more descriptors than the observations, even small perturbations of the training set may lead to large variations in the learning process. In other words, slightly changing the training set results in different linear models and then different sets of nonzero-weighted descriptors. Recent research reported in the literature has shown that if used with care, ensemble modeling can improve the generalization performance particularly for unstable non-linear models, such as those involving neural networks.[28] The technique of bootstrap aggregation (or "bagging")[29,30] is used to stabilize the learning process and ensure that a robust set of features are selected. The idea is to construct a series of individual sparse SVR predictors (models) using a random partition technique[31], record the selected descriptors for each individual bootstrap and then take a union of all descriptors into a single final feature set.

The overall feature selection scheme is illustrated in Figure 2. The following process was carried out in our study:

- Multiple training and validation sets were developed from a master training dataset using a random partition scheme;
- A series of linear SVMs were constructed on the training and validation sets to generate the linear models that achieve good cross-validated correlation coefficients.
- Subsets of features having nonzero weights in the linear models were selected;
- Finally, the features obtained in all previous steps were aggregated to produce the final candidate set of descriptors.

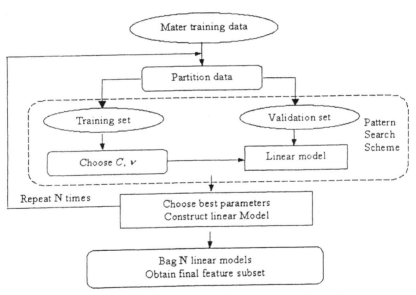

Figure 2. General framework of Feature selection scheme

Nonlinear Regression Bagging Models

Once a set of features is selected, a nonlinear v-SVR with a Gaussian kernel shown in Eq. (6) is used to construct the QSRR models using these features:

$$k(x, x') = \exp\left(-\|x - x'\|^2 / 2\sigma^2\right) \tag{6}$$

This allows us to obtain the regression function f as a linear combination of only a few kernel functions. The sparse nonlinear v-SVR is formulated as follows:

minimize $\quad \dfrac{1}{2}\sum_{j=1}^{M}(\alpha_j + \alpha_j^*) + C\dfrac{1}{M}\sum_{i=1}^{M}(\xi_i + \xi_i^*) + Cv\varepsilon \qquad (7)$

subject to $\quad y_i - \sum_{j=1}^{M}(\alpha_j - \alpha_j^*)\, k(x_i, x_j) - b \le \varepsilon + \xi_i, \quad i = 1,2,\ldots, M$

$\sum_{j=1}^{M}(\alpha_j - \alpha_j^*)\, k(x_i, x_j) + b - y_i \le \varepsilon + \xi_i, \quad i = 1,2,\ldots, M$

$\alpha_j, \alpha_j^*, \xi_i, \xi_i^*, \varepsilon \ge 0, \quad i, j = 1,2,\ldots, M$

The appropriate values for SVR parameters, σ, C and v, were selected by patter search algorithm.[27,32] In order to decrease the possibility if bias on one particular model, the "bagging" technique was again utilized to construct a ensemble of individual SVR predictive models over the selected features, which would be used to generate the final nonlinear SVR predictors $\phi_{bag}(x)$.

$\phi_{bag}(x) = \dfrac{1}{N}\sum_{n=1}^{N}\phi_n(x),$ where N is the cardinality of the ensemble $\qquad (8)$

V. RESULTS AND DISCUSSION

SVR Feature Selection and Bagging Prediction Results

In this feature selection procedure, 20 sparse linear SVM models were constructed based on 20 different random partitions of the training data. In the final aggregate SVR model, there were only seven descriptors remaining with nonzero weights, which would be considered as the most relevant to protein retention response. These seven descriptors and their primary definitions or related chemical information are described in Table 2.

Figure 3 shows protein retention prediction obtained before verse after feature selection based on nonlinear SVR aggregate models. In this figure, the observed retention times (horizontal axis) are plotted against the corresponding predicted values for each protein obtained. The cross validation predictions of training set were shown by blue color, while the blind test data held out during model generation or validation step were indicated by red. The vertical bar shows the full prediction range of retention time of twelve bagged models for each protein and the asterisk in the bar stands for the bagged (average) result of 12 bootstraps for each protein. Before feature selection, the cross-validation for training set produced an $R^2_{CV} = 0.851$ and for the blind test set the bagged result

Table 2: Definition of the relevant descriptors obtained from sparse SVR feature selection

Descriptor Name	Chemical information encoded in these descriptors
PEOE.VSA.FPPOS (MOE)	Fraction of positive polar Van der Waals surface area. The Partial Equalization of Orbital Electronegativities (PEOE) method of calculating the atomic charges was developed by Gasteiger"
FCHARGE (MOE)	Total charge of molecule (sum of formal charges)
PIP2 (TAE)	The second histograni bin of PIP properly. Local average ionization potential in the low range
PIP2O (TAE)	The last histogram bin of PIP property. Local average ionization potential in the high range
SIKIA(TAE)	K electronic kinetic energy density, which correlates with the presence and strength of Bronsted basic sites. (integral average)
SIGIA (TAE)	Derived from the G electronic kinetic energy density on the molecular surface. Similar in interpretation to SIKIA, but provide supplemental information.
VSA.POL	Sum of Van der Waals surface of "polar" atoms

is $R^2_{bag} = 0.926$. On the other way, after feature selection the cross-validated $R^2_{CV} = 0.882$ and the test set $R^2_{bag} = 0.988$. It may be observed that the final non-linear model performs better with only seven features than with the original 243 descriptors. The reduction in features also simplifies the model and allows for better interpretation, which will be discussed in the following paragraphs.

Since the seven descriptors were selected based on an ensemble of SVR models, in the current work, a graphic visualization tool for multivariate data, known as "star plots"[34] in multi-plot format, were used to characterize the relative importance of these descriptors and its consistency across all models based on the size and shape of plots. In our case, each plot consists of a sequence of equi-angular spokes around a given central point representing one descriptor in the investigated data matrix. A line is drawn to connect the ending point of each spoke, producing the star-like appearance. The length of each equi-angular spoke stands for the assigned weight for corresponding descriptor in each of twenty bootstraps or constructed sparse SVR models. Finally, the descriptors were ranked for all 20 bootstrap iterations in column-wise fashion based on either the sum of all 20 radii or the area of the star can be used to represent the overall relative importance of the descriptor over all 20 bootstraps. Descriptors with cyan slash background have negative contributions to the retention time, while the red dot background indicates the positive effect. For instance as shown in following figure, PEOE.VSA.FPPOS has the largest negative effect on retention time and PIP2 has the largest positive effect on retention time.

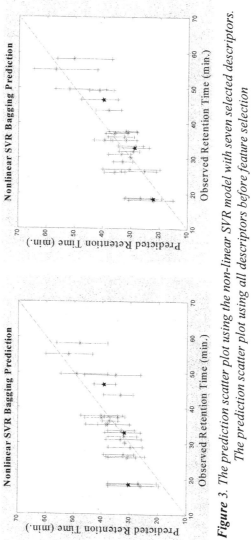

Figure 3. *The prediction scatter plot using the non-linear SVR model with seven selected descriptors. The prediction scatter plot using all descriptors before feature selection*

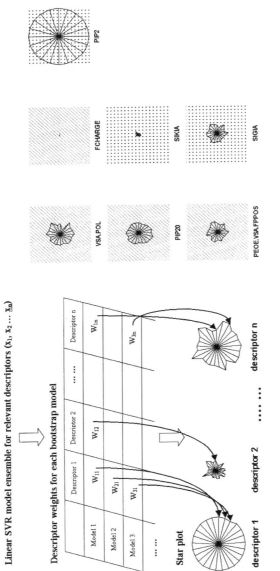

Figure 4. Star plots generation process. Star plots for the seven descriptors selected by the feature selection algorithm.

The MOE descriptor PEOE.VSA.FPPOS represents the fraction of the protein positively charged surface area. As shown in above figure, its negative weight means that greater fractional positive surface area decreases the protein retention time. This result is consistent with the net charge hypothesis that proteins with low negative charge densities will interact more weakly with the resin, and will elute first. It is expected that electrostatic effects between the positively charged functional group on resin surface (quaternary ammonium functional groups $N(CH_3)_3^+$) and the negatively charged protein surfaces will play an important role in anion-exchange system. The same explanation can account for the appearance of another descriptor FCHARGE. This descriptor computes the total formal charge of the protein that is negative in cases where the solution pH is higher than their isoelectric point (P1). The negative sign of its weight in the sparse SVR models is in agreement with the fact that positively charged resin surface exhibit a favorable affinity for the protein with more negative charge. The apparent insignificance of this descriptor is due to the fact that the electrostatic effect has already been represented by other selected electrostatic-related descriptors, such as PEOE.VSA.FPPOS.

MOE descriptor, VSA.POL, approximates the VDW surface area of polar atoms (both hydrogen bond donors and acceptors). The appearance of this descriptor implies that the hydrogen bonding capacity of proteins may also be involved retention process because the surrounding solvent (water) can interact favorably with those charged or polar side chains on the protein surface. Thus, even a protein with a moderate polarity has to pay an energetic penalty that increases in proportion to the overall polar surface of the protein. In other words, a protein with more polar atoms on the exposed van der Waals surface will have a stronger hydrogen-bonding capacity with the mobile phase and will elute out of the column first, accounting for the negative effect of VSA.POJ, for retention shown in the star plot.

Several TAE electron density-based descriptors listed in Table 2 were found to be significant to retention, e.g. SIKIA, SIGIA and PIP. Prior to feature selection, there were twenty PIP descriptors present in the descriptor set, where PIP1 and PIP2 represent regions of the molecular surface where electron density is easily ionizable, while PIP20 is associated with regions of tightly held electron density, such as on exchangeable protons. SIGIA and SIKJA describe the integrals of G and K electronic kinetic energy densities found on the molecular van der Waals surface. These descriptors are associated with the presence and the strength of Lewis basic sites. As shown in star plot, PIP2, SIKIA and SIGIA correlate with increased retention time. This may be due to their representation of increased dipole/induced-dipole or charge/induced-dipole forces between the protein and the strong ion-exchanger groups as well as induced-dipole/induced-dipole interactions between the polarizable aromatic groups of the stationary phase and polarizable regions of the protein. The PIP20 descriptor was found to he anticorrelated with retention time, indicating that the presence of non-acidic hydrogen bond donors (serine, *etc*) increases solute/mobile-phase interactions at the expense of solute/stationary phase interactions.

V. CONCLUSIONS

In this work, Support Vector Machine (SVM) regression methods were introduced in both the feature selection and model construction steps to predict protein retention time in anion-exchange chromatographic systems. Extensive cross-validation of the modeling results was accomplished using bootstrapping approach, and then visualized and interpreted using star plots scheme. This modeling scheme not only has been prove useful for comparative QSRR studies in protein separation studies, but also can be extended to current QSAR study in drug discovery, such as ADME/T virtual screening.

ACKNOWLEDGEMENTS

This work was supported in part by NSF grants IIS-9979860 and BES-0079436. Thanks MOE company for the offering of free MOE license and CCG award for the graduate student support of ACS meeting traveling.

REFERENCES

1. Katritzky, A. R.; Petrukhin, R.; Tatham, D.; Basak, S.; Benfenati. F.; Karelson, M.; Maran, U. *J Chem Inf Comp Sci* **2001**, *41*, 679-685.
2. Montana, M. P.; Pappano, N. B.; Debattista, N. B.; Raba. J.; Luco, J. M. *Chromatographia* **2000**, *51*, 727-735.
3. Mazza, C. B.; Sukumar, N.; Breneman, C. M.; Cramer, S. M. *Anal Chem* **2001**, *73*, 5457-5461.
4. Sutter, J. M.; Peterson, T. A.; Jurs, P. C. *Anal Chim Acta* **1997**, *342*, 113-122.
5. Loukas, Y. L. *J Chromatogr A* **2000**, *904*, 119-129.
6. Kaliszan, R. *Chromatographia* **1977**, *10*, 529-531.
7. Vapnik, V. N. *Ieee T Neural Networ* **1999**, *10*, 988-999.
8. Vapnik, V. N. *The Nature of Statistical Learning Theory*; Springer, Berlin, 1995.
9. Vapnik, V. N. *Estimation of Dependences Based on Emperical Data*; Springer-Verlag, Berlin, 1982.
10. Berman, H. M.; Westbrook, J.; Feng, Z.; Gilliland, G.; Bhat, T. N.; Weissig, H.; Shindyalov, I. N.; Bourne, P. E. *Nucleic Acids Res* **2000**, *28*, 235-242.
11. Breneman, C. M. In *The Application of Charge Density Research to Chemistry and Drug Design*; Piniella, G. A. J. a. J. F., Ed.; Plenum Press, 1991; Vol. NATO ASI Series, pp 357-358.

12. Breneman, C. M.; Thompson, T. R.; Rhem, M.; Dung, M. *Comput Chem* **1995**, *19*, 161-179.
13. Bader, R. F. W. *Atoms in Molecules: A Quantum Theory*; Oxford Univ. Press, Oxford, UK. 1994.
14. Labute, P. *J Mol Graph Model* **2000**, *18*, 464-477.
15. Alex J. Smola, B. S. In *Proccedings ICANN'99, Int. Conf. on Artifitial Neural Networks*; Springer: Berlin, 1999.
16. Scholkopf, B.; Smola, A. J.; Williamson, R. C.; Bartlett, P. L. *Neural Comput* **2000**, *12*, 1207-1245.
17. Whitley, D. C.; Ford, M. G.; Livingstone, D. J. *J Chem Inf Comp Sci* **2000**, *40*, 1160-1168.
18. Sutter, J. M.; Dixon, S. L.; Jurs, P.C. *J Chem Inf Comp Sci* **1995**, *35*, 77-84.
19. Rogers, D.; Hopfinger, A. J. *J Chem Inf Comp Sci* **1994**, *34*, 854-866.
20. So, S. S.; Karplus. M. *J Med Chem* **1996**, *39*, 5246-5256.
21. Zheng, W. F.; Tropsha, A. *J Chem Inf Comp Sci* **2000**, *40*, 185-194.
22. Luke, B. T. *J Chem Inf Comp Sci* **1994**, *34*, 1279-1287.
23. Kubinyi, H. *Quant Struct-Act Rel* **1994**, *13*, 285-294.
24. Izrailev, S.; Agrafiotis, D. *J Chem Inf Comp Sci* **2001**, *41*, 176-180.
25. Izrailev, S.; Agrafiotis, D. K. *Sar Qsar Environ Res* **2002**, *13*, 417-423.
26. Agrafiotis, D. K.; Cedeno, W. *J Med Chem* **2002**, *45*, 1098-1107.
27. Bennett K., B. J., Embrechts M. Breneman C. and Song M. *Journal of Machine Learning Research.* **Submitted**, *1*.
28. Dimitris K. Agrafiotis, W. C., and Victor S. *J Chem Inf Comp Sci* **2002**, *ASAP article*.
29. Breiman, F. *Mach Learn* **1996**, *24*, 123-140.
30. Breiman, F. *Mach Learn* **2001**, *45*, 261-277.
31. Efron, B., Tibshirani, RJ *An introduction to the bootstrap*; Chapman and Hall, New York., 1993.
32. Demiriz A., B. K., Breneman C. and Embrechts M. *Computing Science and Statistics* **2001**.
33. Gasteiger, J.; Marsili, M. *Tetrahedron* **1980**, *36*, 3219-3228.
34. Chambers, J., Cleveland,W., Kleiner,B., Tukey, P. *Graphical Methods for Data-Analysis*; Wadsworth, 1983.

Chapter 9

Analysis of Odor Structure Relationships Using Electronic Van Der Waals Surface Property Descriptors and Genetic Algorithms

Barry K. Lavine[1], Charles E. Davidson[1], Curt Breneman[2], and William Katt[2]

[1]Department of Chemistry, Oklahoma State University, Stillwater, OK 74078-3071
[2]Department of Chemistry, Rensselaer Polytechnic Institute, Troy, NY 12180

This chapter describes a new odor structure relationship (OSR) correlation methodology that utilizes large olfactory databases available in the open scientific literature as input. The first step in this procedure is to represent each molecule in the database by an appropriate set of molecular descriptors. To accomplish this task, Breneman's Transferable Atom Equivalent (TAE) descriptor methodology is used to create a large set of electron density derived shape/property hybrid descriptors. These descriptors have been chosen because they correlate with key modes of intermolecular interactions and contain pertinent information about shape and electronic properties of molecules. In contrast to more traditional methodologies that have shown not to be effective, our use of shape-aware electron density based molecular property descriptors has eliminated many of the problems associated with the use of descriptors based on substructural fragments or chemical topology. A second reason for the limited success of past OSR efforts can be traced to the complex nature of the underlying modeling problem. To meet this challenge, we have developed a genetic algorithm for pattern recognition analysis that selects descriptors, which create class separation in a plot of the two or three largest principal components of the data. Because principal components maximize variance, the bulk of the information encoded by these descriptors is about differences between the odorant classes in a data set.

Olfaction is a poorly understood phenomenon. Although an integral part of every day life, there is a dearth of information about the relationship between chemical structure and odor quality in spite of the fact that many theories have been proposed to correlate molecular structure with the perceived odor quality of a compound [1-7]. It is generally agreed that a compound must be volatile and water and lipid soluble in order for it to have an odor. Beyond this general description, there is no consensus among researchers as to which molecular properties and structural features are responsible for the olfactory impressions invoked by compounds.

Analysis of odor structure relationships (OSRs) using computer generated molecular descriptors and pattern recognition techniques offers a practical approach to the study of odorants. The heart of this approach is finding a set of molecular descriptor from which discriminating relationships can be found. In previously published OSR studies [8-15], only fragment based, topological, and geometric descriptors, (e.g., molecular connectivity indices, substructures, substructural molecular connectivity indices, molecular volume, and principal moments of inertia) have been used to describe molecular shape and characterize the electronic properties of the compounds. Descriptors that contain information about the olfactory process need to be developed and tested in order to formulate more effective OSRs.

We have been studying musk odorants and structurally related nonmusk compounds using computer generated molecular descriptors and pattern recognition techniques. Our interest in musks is both commercial and academic. Almost all fragrances sold commercially contain musks because of their strong fixative properties. Musks are also interesting from an OSR viewpoint because they contain a variety of different structural types. There is a wealth of information in the open scientific literature about musks. Because these compounds have a distinct odor that is rarely confused with any other odor, well characterized data sets with few mislabeled compounds can be obtained from the open literature.

In a previous study [16], we used discriminant analysis to differentiate musk from nonmusk compounds based on a set of 14 computer generated molecular descriptors. A training set consisting of 148 indane, tetralin, and isochroman compounds (67 musks, 81 nonmusks) was studied. A discriminants developed from the set of 14 molecular descriptors correctly assigned every training set compound into its respective category: musk or nonmusk. To test the predictive ability of these descriptors and the linear discriminant associated with them, an external validation set of 15 compounds was used. The same 14 molecular descriptors that correctly classified every compound in the training set were generated for each compound in the validation set. The values of these descriptors were autoscaled using the mean and standard deviation from the original training set data. Of the 15 compounds, 14 were correctly classified by the linear discriminant developed from the training set data. The results of this study indicate that several molecular parameters rather than a single molecular parameter are necessary to predict musk odor quality. Molecular shape was an important factor but it was by no means the only factor for the prediction of musk odor quality. In all likelihood, the perception of odor is probably initiated by the interaction of the odorant with an olfactory receptor site in the nose. Olfactory excitation can only occur if the size and the shape of the stimulant is the complement of the receptor or if the stimulant possesses sufficient conformational flexibility to attain the correct shape. The spatial arrangement of the stimulant's functional and steric groups must also conform to the overall 3-dimensional geometry of the receptor.

In the present study, a new methodology to facilitate the design of new odorants such as musks is described. The introduction of a new odorant can be a lengthy, costly, and laborious process. This process can be streamlined if large olfactory databases available in the open scientific literature are used as input for a new odor structure relationship correlation methodology. The first step in this procedure is to characterize each molecule in the database by an appropriate set of molecular descriptors. To accomplish this task, we have used Breneman's Transferable Atom Equivalent (TAE) descriptor methodology to create a large set of electron density derived shape/property hybrid descriptors [17]. These descriptors have been chosen to represent the problem because they have been shown to be correlated with key modes of intermolecular interactions, and they contain pertinent information about shape and electronic properties of molecules. By comparison, more traditional OSR methodologies have been shown not to be as effective. Our use of shape-aware electron density based molecular property descriptors has eliminated many of the problems associated with the use of descriptors based on substructural fragments or chemical topology.

A second reason for the limited success of past OSR efforts can be traced to the nature of the underlying modeling problem, which often is quite complex. To meet these challenges, we have developed a genetic algorithm for pattern recognition analysis that selects descriptors, which create class separation in a plot of the two or three largest principal components of the data [18-23]. Because principal components maximize variance, the bulk of the information encoded by these descriptors is about differences between the odorant classes in the data set. The principal component analysis routine embedded in the fitness function of the pattern recognition GA acts as an information filter, significantly reducing the size of the search space since it restricts the search to feature sets whose principal component plots show clustering on the basis of class. In addition, the algorithm focuses on those odor classes and/or samples that are difficult to classify as it trains using a form of boosting. Samples that consistently classify correctly are not as heavily weighted as samples that are difficult to classify. Over time, the algorithm learns its optimal parameters in a manner similar to a neural network. The pattern recognition GA integrates aspects of artificial intelligence and evolutionary computations to yield a smart one pass procedure for feature selection and classification. The efficacy of this methodology has been evaluated using a structurally diverse database consisting of 331 macrocyclic and nitroaromatic compounds (192 musks and 139 nonmusks).

Musk Data Set

All compounds used in the present study have been taken from literature reports of chemical structure and odor quality [24-29]. The structural classes present in the data set are shown in Figure 1. Natural musks, whose sources include both rare animal and plant species, are macrocycles. The first synthetic musks prepared were nitrated derivatives of benzene. The 192 macrocyclic and nitroaromatic musks are of strong, medium, weak, or unspecified odor intensity, whereas the 139 nonmusks are odorless or have an odor other than musk. We deliberately chose the nonmusks to be as similar in structure to the musks as possible. This not only adds the extra challenge of separating very similar structures on the basis of odor quality, but also increases our understanding of how small structural changes can affect odor character. Further details about this data set can be found elsewhere [30].

Each compound in the data set was characterized by a set of computer generated molecular descriptors. The present work emphasizes the use of electron density derived descriptors of three general types: TAE or molecular surface property descriptors, surface property wavelet coefficient descriptors (WCD), and PEST surface property hybrid descriptors. This methodology takes advantage of a new, rapid, charge density reconstruction algorithm that employs atomic charge density fragments that have been pre-computed using *ab initio* wave functions. A library of atomic charge density components is used to construct molecular electronic densities in a form that allows for rapid retrieval of the molecular surface properties needed to generate descriptors. For each calculated molecule, the program reads in molecular structure information, and then reconstructs the electronic properties of the molecular surface from the atomic fragments. The distributions of several electronic properties on molecular surfaces may then be quantified to give a large variety of numerical descriptors. The CPU and disk resources required for these calculations are minimal.

The underlying methodology relies on the hypothesis that a causative relationship exists between observed odor properties and the distribution of certain molecular electronic properties as sampled on molecular van der Waals surfaces. An additional hypothesis for PEST shape/property hybrid descriptor validation is that spatial arrangements of surface electronic properties contain pertinent chemical information. Both of these hypotheses have been previously validated [31] in studies involving biological and nonbiological molecular behavior.

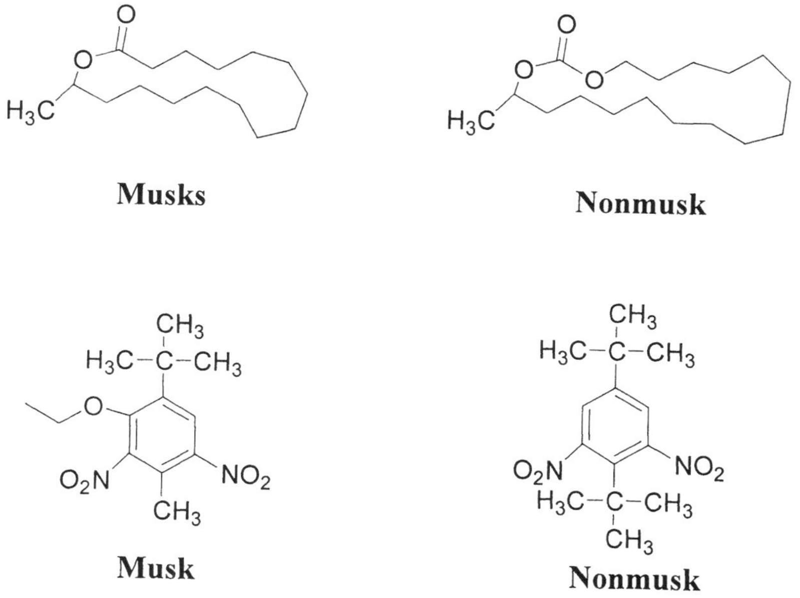

Musks

Nonmusk

Musk

Nonmusk

Figure 1. Two strong musks and two nonmusks representing the major structural classes of compounds found in the data set. (Reproduced from Reference 30. Copyright 2003 American Chemical Society.)

Pattern Recognition Analysis

For pattern recognition analysis, each compound was initially represented by 896 computer generated molecular descriptors derived from connection tables or from three dimensional models of the compounds. The connection tables and the three dimensional models of the compounds were generated by the modeling program Quanta (Molecular Simulations), which contained a molecular mechanics model building routine that utilized the CHARMN force field. Traditional molecular property descriptors, including connectivity based topological 2D descriptors and physiochemical property descriptors were computed via the MOE Program (Chemical Computing Group, Montreal, Canada) and were included in the study. TAE molecular surface property reconstructions were generated using Convert2001 [32]. The Property Encoded Surface Translator (PEST) algorithm [33] was used to generate wavelet and hybrid shape/property descriptors. Molecular electron density properties for all compounds in the study were represented by TAE surface histogram descriptors, wavelet coefficient descriptors, and hybrid shape/property descriptors.

Before any given descriptor was entered into the study, it was checked to see whether it had the same value for all compounds in the training set. A descriptor would be eliminated from consideration if it was invariant. Prior to pattern recognition analysis, each descriptor was autoscaled to zero mean and unit standard deviation to alleviate any problems arising from scaling and facilitate the identification of informative descriptors.

The premise underlying the approach to pattern recognition used in the present study is that all data analysis methods will work well when the problem is simple. By identifying the appropriate features, a "hard" problem can be reduced to a "simple" one. Therefore, our goal is feature selection. To ensure identification of all relevant descriptors, it is best that a multivariate approach to feature selection be employed. The approach should also take into account the existence of redundancies in the data.

A genetic algorithm (GA) for pattern recognition analysis was used to identify molecular descriptors from which discriminating relationships could be found. The pattern recognition GA selects descriptors that optimize the separation of the classes in a plot of the two or three largest principal components of the data. The principal component analysis routine embedded in the fitness function of the GA acts as an information filter, significantly reducing the size of the search space, since it restricts the search to features whose principal component plots show clustering on the basis of class. In addition, the algorithm focuses on those classes and or samples that are difficult to classify as it trains using a form of boosting to modify the class and sample weights. Samples that consistently classify correctly are not as heavily weighted as those samples that are difficult to classify. Over time, the algorithm learns its optimal parameters in a manner similar to a neural network.

To facilitate the tracking and scoring of the principal component plots, class and sample weights, which are an integral part of the fitness function, are computed (see equations 1 and 2) where CW(c) is the weight of class c (with c varying from 1 to the total number of classes in the data set). $SW_c(s)$ is the weight of sample s in class c. The class weights sum to 100, and the sample weights for the objects comprising a particular class sum to a value equal to the class weight of the class in question.

$$CW(c) = 100 \frac{CW(c)}{\sum_c CW(c)} \quad (1)$$

$$SW(s) = CW(c) \frac{SW(s)}{\sum_{s \in c} SW(s)} \quad (2)$$

The principal component plot generated for each feature subset after it has been extracted from its chromosome is scored using the K-nearest neighbor classification algorithm [34]. For a given data point, Euclidean distances are computed between it and every other point in the principal component plot. These distances are arranged from the smallest to largest. A poll is then taken of the point's K_c nearest neighbors. For the most rigorous classification, K_c equals the number of samples in the class to which the point belongs. (K_c often has a different value for each class.) The number of K_c nearest neighbors with the same class label as the sample point in question, the so-called sample hit count, SHC(s), is computed ($0 \leq SHC(s) \leq K_c$) for each sample. It is then a simple matter to score a principal component plot (see equation 3). First, the contribution of each compound in class 1 to the overall fitness function is computed, with the scores of all the compounds comprising the class summed to yield the contribution by this class to the overall fitness. This same calculation is repeated for the other class with the scores from each class summed to yield the overall fitness, F (d).

$$F(d) = \sum_c \sum_{sec} \frac{1}{K_c} \times SHC(s) \times SW(s) \quad (3)$$

To understand scoring, consider a data set with two classes, which have been assigned equal weights. Class 1 has 10 compounds, and class 2 has 20 compounds. For uniformly distributed sample weights, class 1 compounds will have a weight of 5 and class 2 compounds will have a weight of 2.5, since each class has a weight of 50 and the sample weights in each class are uniformly distributed. Suppose a compound in class 1 has, as its nearest neighbors, 7 class 1 compounds in a principal component plot developed from a particular feature subset. Hence, $SHC(c)/K_c = 7/10$, and the contribution of this compound to the fitness function for the particular feature subset equals 0.7*5 or 3.5. Multiplying SHC/K_c by $SW(s)$ for each compound and summing up the corresponding product for the 30 compounds in the data set yields the value of the fitness function for this particular set of features.

The fitness function of the pattern recognition GA is able to focus on individual compounds and classes of compounds that are difficult to classify by boosting their weights over successive generations. In order to boost, it is necessary to compute both the sample-hit rate (SHR), which is the mean value of SHC/K_c over all feature subsets produced in a particular generation (see equation 4), and the class-hit rate (CHR), which is the mean sample hit rate of all samples in a class (see equation 5). ϕ in equation 5 is the number of chromosomes in the population, and AVG in equation 6 refers to the average or mean value. During each generation, class and sample weights are adjusted by a perceptron (see equations 6 and 7) with the momentum, P, set by the user. (g + 1 is the current generation, whereas g is the previous generation.) Classes with a lower class hit rate are more heavily boosted than those classes that score well.

$$SHR(s) = \frac{1}{\phi} \sum_{i=1}^{\phi} \frac{SHC_i(s)}{K_c} \quad (4)$$

$$CHR_g(c) = AVG(SHR_g(s) : \forall_{sec}) \quad (5)$$

$$CW_{g+1}(s) = CW_g(s) + P(1 - CHR_g(s)) \quad (6)$$

$$SW_{g+1}(s) = SW_g(s) + P(1 - SHR_g(s)) \quad (7)$$

Boosting is crucial for the successful operation of the pattern recognition GA because it modifies the fitness landscape by adjusting the values of both the class and sample weights. This helps to minimize the problem of convergence to a local optimum. Hence, the fitness function of the GA changes as the population is evolving towards a solution.

During each generation, selection, crossover, and mutation operators are applied to the chromosomes. Fit strings are retained and selected for breeding, a process called selection. The fit feature subsets are broken-up, swapped, and recombined, creating new feature subsets, which are introduced into the population of potential solutions. This process is called crossover. In this study, the selection and crossover operators are implemented by ordering the population of strings, i.e. potential solutions, from best to worst, while simultaneously generating a copy of the same population and randomizing the order of the strings in this copy with respect to their fitness. A fraction of the population is then selected as per the selection pressure which is usually set at 0.5. The top half of the ordered population is mated with strings from the top half of the random population, guaranteeing the best 50% are selected for reproduction, while every string in the randomized copy has a uniform chance of being selected. If we used a purely biased selection criterion to select strings, only a small region of the search space would be explored. Within a few generations, the population would consist of only copies of the best strings from the initial population.

For each pair of strings selected for mating, two new strings are generated using three-point crossover In the last step of reproduction, a mutation operator is applied to the new strings. The mutation probability of the operator is usually set at 0.01, so 1% of the feature subsets are selected at random for mutation. A chromosome marked for mutation has a single random bit flipped, which allows the GA to explore other regions of the parameter space. If the GA finds a better solution, the genes from this point can invade the population, with the optimization continuing in a new direction.

The resulting population of strings, both parents and children, are sorted by fitness, and the top ϕ strings are retained for the next generation. Because the selection criterion used for reproduction exhibits bias for the higher-ranking strings, the new population is expected to perform better on average than its predecessor. The reproductive operators, however, also assure a significant degree of diversity in the population, since the crossover points and reordering of exchanged string fragments of each chromosome pair is selected at random. The aforementioned procedure, which involves evaluation, reproduction, and boosting of potential solutions, is repeated until a specified number of generations are executed or a feasible solution is found. A block diagram of the genetic algorithm

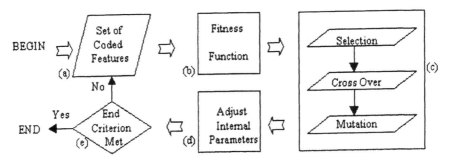

Figure 2. Block diagram of the genetic algorithm developed for pattern recognition analysis.

(GA) developed for pattern recognition is shown in Figure 2. Boosting is denoted by the block entitled, "adjusting internal parameters," in the figure.

Results and Discussion

The first step in the study was to apply principal component analysis (PCA) to the training set data. PCA is the most widely used multivariate analysis technique in science and engineering. It is a method for transforming the original measurement variables (i.e., the molecular descriptors) into new uncorrelated variables called principal components. Each principal component is a linear combination of the original measurement variables. Using this procedure, a new coordinate system is developed that is better at conveying the information present in the data than axes defined by the original measurement variables. This new coordinate system is linked to variance. Often, only two or three principal components are necessary to explain all of the information present in the data if there are a large number of interrelated measurement variables. Hence, principal component analysis is often applied to multivariate data for dimensionality reduction, in order to identify outliers, display structure, and classify samples.

Figure 3 shows the results of a principal component mapping experiment for the 331 training set compounds and the 871 TAE and MOE derived descriptors. The macrocyclic nonmusks are denoted by the symbol 1, the aromatic nitro nonmusks are denoted by the symbol 2, macrocyclic musks are denoted by the symbol 3, and the aromatic nitro musks are denoted by the symbol 4. The two largest principal components of the data explain 35% of the total cumulative variance of the data. From the plot it is evident that most of the information captured by the two largest principal components is about chemical structure since the macrocycles are well separated from the nitro aromatics on the first principal component.

We used the pattern recognition GA to identify a subset of the 871 descriptors from which a discriminating relationship could be developed for the macrocyclic and nitroaromatic musks. These molecular descriptors were identified by sampling key descriptor subsets, scoring their principal component plots, and tracking those compounds that were difficult to classify. After 100 generations, the pattern recognition GA identified 16 TAE derived descriptors whose principal component plot (see Figure 4) showed clustering of the compounds on the basis of odor quality. The principal component plot explains 57% of the total cumulative variance of the data. Because the musks are well separated from the nonmusks on the first principal component, one can conclude that musk odor activity can be accurately modeled by TAE derived descriptors. Furthermore, TAE derived descriptors can span structural manifolds.

The 16 molecular descriptors identified identified by the pattern recognition GA are listed in Table 1. Most of the 16 descriptors identified by the pattern recognition GA convey information about intermolecular interactions which suggests their importance in defining musk odor quality. DGNH8, DGNW7, and DGNW21 are correlated to weak bonding interactions and probably describe some facet of the interaction between the musk and the receptor. DGNB52, a shape descriptor, can be interpreted to mean that it is probably crucial for areas with a low rate of change in the G kinetic energy to be relatively far apart in the molecule. LAPLB30, another shape descriptor, is important in characterizing donor/acceptor relationships.

DKNMAX is correlated to hydrophobicity and polarizability. DKNW6 is a wavelet descriptor emphasizing the same properties encoded in DKNMAX, whereas DKNB55 is a shape descriptor developed from the rate of change of the K kinetic energy density. Evidently, long distances between relatively high values of the rate of change in the K kinetic energy density is crucial for musk odor quality. Both DRNH6 and DRNW26 are highly correlated to the four DG descriptors discussed previously. PIPH8, PIPW18, and PIPW29 are descriptors that convey information about the local ionization potential of the molecule. BNPW4 and BNPW9 are so-called bare nuclear potential descriptors, which probably describe interactions involving polar and hydrogen bonding. ANGLEB52 is a special type of shape descriptor. Molecules that are curved and do not have sharp turns are favored by this descriptor.

Figure 4 shows that nitrated and nitro-free musks have common structural features that can be used to differentiate them from nonmusks. This is a significant result. Fragrance chemists have long sought to discover the overlap between nitrated and nitro free musks in terms of the

138

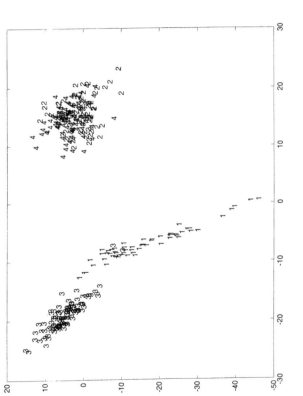

Figure 3. A plot of the two largest principal component plot of the 331 compounds and the 871 TAE and MOE descriptors comprising the musk database. 1 = macrocyclic nonmusk, 2 = aromatic nitro nonmusks, 3 = macrocyclic musks, and 4 = aromatic nitro musks. (Reproduced from Reference 30. Copyright 2003 American Chemical Society.)

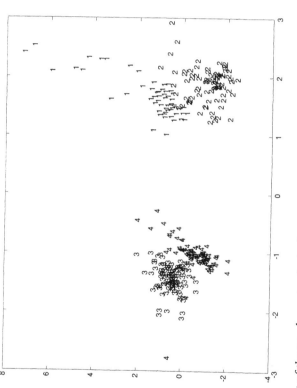

Figure 4. A plot of the two largest principal component plot of the 331 compounds and 16 TAE derived descriptors identified by the pattern recognition GA. 1 = macrocyclic nonmusk, 2 = aromatic nitro nonmusks, 3 = macrocyclic musks, and 4 = aromatic nitro musks. (Reproduced from Reference 30. Copyright 2003 American Chemical Society.)

Table 1. Descriptors Selected by the Pattern Recognition GA

DGNH8	Rate of change of the G kinetic energy density normal to and away from the surface. The "H" defines this descriptor as a histogram descriptor and the "8" means that its value is from the 8th bin. Evidently, large values of DGN are important for musk odor quality.
DGNW7	Scale wavelet descriptor ("W") describing the same basic value type as DGNH8.
DGNW21	Detail coefficient wavelet descriptor ("W") describing the same basic value type as DGNH8.
DGNB52	Shape descriptor. The "5" indicates that long rays are important and the "2" indicates that low DGN values are important.
LAPLB30	The Laplacian is the second derivative of the electronic energy distribution. It is a shape descriptor with "3" meaning that intermediate length rays are represented, and "0" meaning that small property values are represented.
DKNMAX	Maximum value of DKN. It is similar to DGNMAX except that it uses K kinetic energy density, which is often complementary to the G kinetic energy density.
DKNW6	Scale wavelet descriptor emphasizing the importance of changes in the K kinetic energy.
DKNB55	Shape descriptor representing relatively long rays between relatively high values of DKN.
DRNH6	This descriptor represents the rate of fall off of the electron density. The "H" defines this descriptor as a histogram descriptor and the "6" means that its value is from the 6th bin. This type of descriptor is often highly correlated to DGN.
DRNW26	A detail wavelet descriptor for DRN.
PIPH8	PIP is the local average ionization potential. The "8" represents a bin containing lower than average values for PIP.
PIPW18	Detail coefficient wavelet descriptor for PIP
PIPW29	Detail coefficient wavelet descriptor for PIP
BNPW4	Scale coefficient wavelet descriptor of the bare nuclear potential. It is suspected of describing polar and hydrogen bonding interactions.
BNPW9	Scale coefficient wavelet descriptor of the bare nuclear potential. It is suspected of describing polar and hydrogen bonding interactions.
ANGLEB52	Special shape descriptor. For long rays (5), the angle (2) is less than average.

structural features that a compound must possess in order to evoke a musky odor. According to the most recent theory of olfaction, compounds interact with multiple olfactory receptors rather than individual ones. Because the olfactory receptors are not very selective towards specific odorants, identification of an odor is based on a distinct pattern of responses. It is, therefore, plausible that nitro musks such as "musk ketone" and "musk ambrette" long used in the fragrance industry because of their fixatives properties generate a response pattern, which in some measure overlaps with the response pattern generated by nitro-free musks. This viewpoint would be consistent with the SAR results obtained in this study and would also validate a viewpoint long held by perfumers that nitrated musks fit better in the musk odor category than in any other odor category even though nitrated musks elicit an odor response that is markedly different from nitro free musks.

The SAR of aromatic nitro musks is not well understood because of the complex substitution pattern and the varied polyfunctional character of the nitro group. Aromatic nitro musks have highly impure and informationally complex odors. Nevertheless, aromatic nitro musks could be separated from nonmusks by a single principal component developed from 16 TAE derived descriptors. Clearly, the musk odor activity of aromatic nitro musks can be accurately modeled by TAE derived descriptors. In our opinion, this constitutes an important step forward in the study of olfactory relationships since it has been demonstrated via macrocylic and aromatic nitro musks that TAE derived descriptors convey significant information about molecular interactions that are important in olfaction.

References

1. Wright, R. H. "Odor and Molecular Vibration: Neural Coding of Olfactory Information," *J. Theor. Biol.*, **1977**, 64, 473-502.
2. Turin, L. "A Spectroscopic Mechanism for Primary Olfactory Reception," *Chem. Senses,* **1996**, 21, 773-791.
3. Beets, M. G. J. *Molecular Structure and Organoleptic Quality*; Macmillan: NY 1957.
4. Theimer, E.; J. T. Davies. "Olfaction, Musk Odor, and Molecular Properties," *J. Agric. Food. Chem.*, **1967**, 15(1), 6-14.
5. Amoore, J. E. *Molecular Basis of Odor*; Charles C. Thomas: Springfield, IL, 1970.
6. Dravnieks, A.; Laffort, P. "Physico-chemical Basis of Quantitative and Qualitative Odor Discrimination in Humans," in *Olfaction and Taste*; Schneider, D. (Ed.); Wissens-Verlag-MBH: Stuttgart, FRG, 1972; pp. 142-148.

7. Axel, R. "The Molecular Logic of Smell," *Scientific American*, October **1995**, 154-159.
8. Brugger, W. E.; Jurs, P. C., "Extraction of Important Molecular Features of Musk Compounds Using Pattern Recognition Techniques," *J. Agric. Food Chemistry*, **1977**, 25(5), 1158-1164.
9. Klopman, G.; Ptscelintsev, D. "Application of the Computer Automated Structure Evaluation Methodology to a QSAR Study of Chemoreception. Aromatic Musky Odorants," *J. Agric. Food Chem.*, **1992**, 40, 2244-2251.
10. Kier, L. B.; Di Paolo, T.; Hall, L. H. "Structure Activity Studies on Odor Molecules Using Molecular Connectivity," *J. Theor. Biol.* **1977**, 67, 585-596.
11. Greenberg, M. J. "Dependence of Odor Intensity on the Hydrophobic Properties of Molecules. A Quantitative Structure Odor Intensity Relationship," *J. Agric. Food Chem.*, **1979**, 27, 347-352.
12. Wolkowski, Z. W.; Moccatti, D.; Heymans, F.; Godfroid, J. J. "A Quantitative Structure-Activity Approach to Chemoreception, Importance of Lipophilic Properties," *J. Theor. Biol.*, **1977**, 66, 181-193.
13. Rossiter, K. J., "Structure-Odor Relationships," *Chem. Rev.*, **1996**, 96, 3201-3240.
14. G. Frater, J. A. Bajgrowicz, and P. Kraft, "Fragrance Chemistry," *Tetrahedron,* **1998**, 54, 7633-7703.
15. D. Cherqaoui, M'Hamed Esseffar, D. Villemin, J.-M. Cense, M. Chastrette, and D. Zakarya, "Structure-Musk Odor Relationship of Tetralin and Indan Compounds Using Neural Networks," *New Journal of Chemistry*, **1998**, 22, 839-843.
16. Narvaez, J.N.; Lavine, B. K.; Jurs, P. C. "Structure-Activity Studies of Musk Odorants Using Pattern Recognition: Bicyclo and Tricyclo-Benzenoids," *Chemical Senses*, **1986**, 11(1), 145-156.
17. Breneman, C. M.; Thompson, T. R.; Rhem, M.; Dung, M. "Electron Density Modeling of Large Systems Using the Transferable Atom Equivalent Method," *Comput. Chem.*, **1995**, 19(3), 161-172.
18. Lavine, B. K.; Davidson, C. E.; Rayens, W. T. "Machine Learning Based Pattern Recognition Applied to Microarray Data, **Combinatorial Chemistry & High Throughput Screening**," 2004, 7, 115-131.
19. Lavine, B. K.; Davidson, C. E.; Vander Meer, Robert K.; Lahav, S.; Soroker, V.; Hefetz, A. "Genetic Algorithms for Deciphering the Complex Chemosensory Code of Social Insects," **Chemometrics & Intelligent Laboratory Instrumentation**, 2003, 66(1), 51-62.
20. Lavine, B. K.; Davidson, C. E.; Moores, A.J. "Genetic Algorithms for Spectral Pattern Recognition," Vibrational Spectroscopy, 2002, 28(1), 83-95.

21. Lavine, B. K.; Davidson, C. E.; Moores, A. J. "Innovative Genetic Algorithms for Chemoinformatics, " **Chemometrics & Intelligent Laboratory Instrumentation**, 2002, 60(1), 161-171.
22. Lavine, B. K.; Vesanen, A.; Brzozowski, D.; Mayfield, H. T. "Authentication of Fuel Standards using Gas Chromatography/Pattern Recognition Techniques," **Anal Letters**, 2001, 34(2), 281-294.
23. Lavine, B. K.; Davidson, C. E.; Moores, A. J.; Griffiths, P. R. "Raman Spectroscopy and Genetic Algorithms for the Classification of Wood Types," **Applied Spectroscopy**, 2001, 55(8), 960-966.
24. Beets, M. G. J. *Structure Activity Relationships in Human Chemoreception*; Applied Science Publishers: London, 1978.
25. Wood, T. F. "Chemistry of the Aromatic Musks," *Givaudanian*, Clifton, NJ. **1970**, pp. 1-37.
26. Bersuker, I. B.; Dimoglo, A. S.; Gorbachov, M. Yu; Vlad, P. F.; Pesaro, M. "Origin of Musk Fragrance Activity: The Electron-Topological Approach," *New J. Chem.*, **1991**, 15, 307-320.
27. MacLeod, A. J. in *Olfaction in Mammals, Symposia of the Zoological Society of London,* No. 45, Stoddard, D.M. Ed.; Academic Press: New York, 1980; p. 15.
28. Theimer, E. T. (Editor). *Fragrance Chemistry, The Science of the Sense of Smell*; Academic Press: New York, 1982.
29. Ham, C. L.; Jurs, P. C. "Structure Activity Studies of Musk Odorants Using Pattern Recognition: Monocyclic Nitrobenzenes," *Chemical Senses*, **1985**, 10(4), 491-502.
30. Lavine, B. K.; Davidson, C. E.; Breneman, C.; Katt, W. "Electronic Van der Waals Surface Property Descriptors and Genetic Algorithms for Developing Structure-Activity Correlations in Olfactory Databases," **J. Chem. Inf. Science,** 2003, 43, 1890-1905.
31. Breneman, C. M.; Rhem, M. "QSPR Analysis of HPLC Column Capacity Factors for a Set of High-Energy Materials Using Electronic van der Waals Surface Property Descriptors Computed by Transferable Atom Equivalent Method," *J. Comput. Chem.*, **1997**, 18(2), 182-197.
32. Song, M.; Breneman, C. M.; Bi, J.; Sukumar, N.; Bennett, K. P.; Cramer, S.; Tugcu, N. "Prediction of Protein Retention Times in Anion-Exchange Chromatography Systems Using Support Vector Regression," *J. Chem. Inf. Comput. Sci.*, **2002**, 42(6), 1347-1357.
33. Breneman, C.; Bennett, Bi, J.; Song, M.; Embrechts, M. "New Electron Density-Derived Descriptors and Machine Learning Techniques for Computational ADME and Molecular Design," MidAtlantic Computational Chemistry Meeting, Princeton University, Princeton, NJ, April 17, **2002**.
34. James, M. Classification: John Wiley & Sons: New York, **1992**.

Chapter 10

Optimization of MDL Substructure Search Keys for the Prediction of Activity and Toxicity

Douglas R. Henry and Joseph L. Durant, Jr.

MDL, 14600 Catalina Street, San Leandro, CA 94577

This article describes the algorithmic generation of MDL substructure search keys and the optimization of the key definitions and weightings for predicting the biological activity of chemical structures. Substructure search keys are bitsets representing functional groups and atom pairs in molecules. They are mainly used for molecular similarity calculations, but they are also useful for clustering and classification analysis. We applied genetic algorithms to generate keysets that could better predict activity in a set of structures first used by Briem and Lessel (Briem, H. and Lessel, U., *Perspect. Drug Discov. Design*, 2000, **20**, 231-244). Prediction performance improved from 65% to 74% correctly classified using a 324-keyset. We then applied a variety of weighting schemes in similarity calculations to discriminate between drug structures from the MDL Drug Data Report (MDDR) database and toxic structures from the MDL Toxicity database. The best results were obtained when the keys were weighted according to the inverse of database frequency (79% correct), followed by surprisal and unit weighting. Using coefficients from principal component and discriminant analyses did not yield better results.

In the field of drug discovery perhaps the most common pursuit is attempting to predict the biological activity of a molecule from its chemical structure. We have not yet reached the point where structures *per se*, can be used as "data" although virtual screening and docking approach this goal. It is still customary to convert structures to some numerical quantities (descriptors) and by the "similar property principle" we expect compounds that are similar structurally to have similar descriptor values and to some degree, similar chemical and biological properties (1). The numbers and types of descriptors used to encode chemical structure information are huge. One descriptor calculation program, DRAGON, can compute nearly a thousand descriptors (2). Although there are many ways to organize and classify such descriptors, a major distinction exists between descriptors that attempt to encode molecular properties, often sensitive to 3D shape or conformation (e.g., molecular shape, Log *P*, polar surface area), and descriptors that encode purely structural information, typically based on the 2D connection table of the structure (e.g., substituent constants and fragment or substructure keys). The "dimensionality" of chemical *property* space is finite and limited. For example, in the field of quantitative structure-activity relationships (QSAR) it is common to speak of steric, electronic, and lipophilic properties as being the most important for drug binding and activity. As a consequence, we often find a high degree of correlation among the multitude of chemical property descriptors. The dimensionality of chemical *structure* space is much greater, although as Bemis and co-workers have shown, for drug structures at least, it is still somewhat limited (3, 4). Accordingly, we would expect a larger number of descriptors to be required to adequately cover chemical structure space.

A common type of structure-based topological descriptor is the substructure or fragment bitset, sometimes referred to as substructure "keys" or "fingerprints". In this type of binary descriptor a set of bits – usually hundreds or thousands of bits long – encode the presence of specific functional groups or atom/bond combinations in a structure. With the introduction of chemical structure databases and substructure searching in the early 1980's, substructure keys were originally used as filters to aid the search process. Thus, if a query substructure has a carbonyl group, every structure containing that substructure must also have a carbonyl group. A particular bit in a set of substructure keys can be checked using fast binary logic to filter out structures in the database that lack the required functionality. As chemical database systems moved to more sophisticated tree-based indexing of structures, substructure keys were no longer needed for substructure searching (5).

A new use for substructure keys emerged in the 1990's with the emphasis on combinatorial chemistry, library design, and chemical diversity. A simple and chemically intuitive measure of the similarity between two structures can be obtained using one of several binary similarity coefficients (6). The most common of these is the Tanimoto coefficient, which is the defined as the number of bits set in common between two structures, divided by the total number of bits set between them (i.e., the intersection divided by the union of the two sets of bits). This coeffieicnt ranges in value from 0 for no similarity to 1.0 for perfect overlap:

$$S_{a,b} = \frac{\sum bits\_in\_common_{a,b}}{\sum bits\_set_a + \sum bits\_set_b - \sum bits\_in\_common_{a,b}} \quad (1)$$

The Tanimoto coefficient computes similarity considering only the presence of features, not their absence. In addition, it typically weights all features equally. In a database of structures, a common feature like a phenyl ring, may not differentiate structures as well as a less common feature like a cyclopropyl group. For this reason, MDL database programs such as ISIS™ commonly compute a weighted Tanimoto similarity coefficient:

$$S_{a,b} = \frac{\sum bits\_in\_common_{a,b} \cdot weights_{a,b}}{\sum bits\_set_a \cdot weights_a + \sum bits\_set_b \cdot weights_b - \sum bits\_in\_common_{a,b} \cdot weights_{a,b}} \quad (2)$$

For any given key, the weight is usually calculated to be inversely proportional to the frequency of occurrence of the feature in the database. It is possible to recalculate weights from the database automatically or to assign weights manually. The values of a weight typically range from 0 to 100.

Several groups have studied the use of MDL keys in the past, mainly as they relate to molecular diversity. Brown and Martin (7) and McGregor and Pallai (8) studied clustering and diversity. Combinatorial library generation and evaluation was studied by Brown and Martin (9), Koehler et. al. (10), and Ajay et. al. (11). Information content and structure comparison was the focus of articles by Brown and Martin (12), Jamois et. al. (13), and Briem and Lessel (14).

Defining MDL Keys

Two main approaches to the design of substructure search and similarity keys are found in commercial chemical databases. In one approach, offered by Daylight Chemical Information Systems, the functional groups are "discovered" algorithmically in the database by computing all paths in the molecules of the database up to a given bond distance, and assigning each unique path to a key in the bitset. This can yield large numbers of sparsely-populated keys, so typically the keys are "folded" to increase the density of the bits that are set. This means the bit string is split in half, and the second half is logically OR'ed with the first half. This process is repeated until the required density of bits is achieved. Efficiency of the keys is optimal when about half of the bits are set on average, per structure. An alternative approach, followed by MDL and other software vendors, is to define a fixed number of keys, using specific structural features for each key. In MDL databases, two keysets have been available for molecules (as opposed to reactions and 3D models, which have their own keysets). These are the 960-key SSKEYS set and the 166-key USERKEY set. Each molecule that gets registered into an ISIS database has the keys calculated as part of the registration process, and stored in the database for similarity searching purposes. Keys in the 960-keyset are defined in a table or file known as the ENKFIL. Figure 1 shows an example of one line in the ENKFIL, along with the interpretation of the various entries in the line, and a graphical representation of a substructure corresponding to the given key. The 960-keyset encodes 1387 possible atom-bond combinations (substructures), with additional keys for multiple occurrences. The keys are "confounded" such that a given substructure can set up to three keys, and a given key may be set by multiple, different substructures (typically ones that are orthogonal in the database – they would not usually occur together in a given structure).

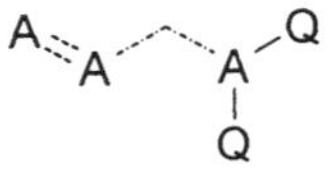

"2 3 5 2 3 1 479 469 763"

(2) 2-bond distance between atoms with:
  (3) - a multiple, non-aromatic bond
  (5) - at least two heteroatom neighbors

(2) 2 or more occurrences of the substructure

(3) the descriptor sets 3 keybits

(1) these keybits can be set by other descriptors

(479...) the keybits set are 479, 469, and 763

Figure 1. Typical entry in the ENKFIL table to define a substructure key. The substructure corresponding to the given key is shown (A=any atom type, Q=hetero atom type, and dot-dash bond=any bond type).

149

The 166-key subset was originally developed to encode common substructural features found in organic molecules. For this reason, this keyset is typically used for filtering in structure searches or it is the set most often used in diversity and biological activity prediction studies. The substructures are more intuitive than many of those present in the 960-keyset. Table I shows the first 24 of the 166-keyset definitions. Each key in the 166-keyset maps to one of the 960 keys as well, though not in any particular order. A complete description of the definition of the MDL keys is found in Durant et. al. (15).

Table I. Key definitions for the first 24 keys in the MDL 166-keyset.

Key	Definition	960-key Equivalent
1	ISOTOPE	1
2	103 < ATOMIC NO. < 256	2
3	GROUP IVA,VA,VIA PERIODS 4-6 (GE..)	29
4	ACTINIDE	4
5	GROUP IIIB,IVB (SC..)	6
6	LANTHANIDE	7
7	GROUP VB,VIB,VIIB (V..)	8
8	QAAA@1	9
9	GROUP VIII (FE..)	10
10	GROUP IIA (ALKALINE EARTH)	12
11	4M RING	14
12	GROUP IB,IIB (CU..)	15
13	ON(C)C	16
14	S-S	17
15	OC(O)O	18
16	QAA@1	19
17	CTC	22
18	GROUP IIIA (B..)	23
19	7M RING	27
20	SI	28
21	C=C(Q)Q	30
22	3M RING	32
23	NC(O)O	42
24	N-O	46

Constructing "Better" Keysets

The 960 and 166-keysets have been used in MDL databases for many years without any changes. They were originally designed to filter structures and to aid substructure searching. It is rather remarkable that they would also be found useful for the purpose of predicting biological activity. It was natural to ask whether improved biological prediction could be obtained by modifying the keys, either by redefining them, or by adjusting their weights. Accordingly, we did experiments to try to optimize 1) the key definitions, 2) the total number of keys, and 3) the key weights. The first step was to pick an objective criterion to measure the effectiveness of the keys for predicting biological activity. Briem and Lessel (14) studied a set of 383 compounds with known activity, plus a set of 574 compounds of "random" activity, all selected from the MDL Drug Data Report database (MDDR) (16). They computed the similarity of each structure to all the others, using a variety of fingerprint descriptors, including MDL keys. For each structure, they examined the ten most similar neighbors and computed the fraction of those that had the same activity class as the given structure. The average percentage "correct" by this calculation was used as a measure of the effectiveness of the given type of descriptor for predicting biological activity.

To study changes in the definition and selection of keys we adopted the Briem and Lessel success measure, with an added correction in the case of ties (15). A total of five sets of keys were generated and examined:

1. The standard MDL 960-keyset.
2. The standard MDL 166-keyset.
3. A 726-keyset obtained by removing multiply-mapped keys from the 960-keyset (i.e., removing keys that mapped to several substructures).
4. A 1387-keyset obtained by generating singly-encoded keys from the multiply-mapped keys for the 960-keyset (i.e., if several substructures mapped to a given key, they were now mapped each to a separate key).
5. A 3234-keyset obtained by encoding every possible combination of the atom and bond/path properties used in the 960-keyset.

The first question we asked was "How many keys are really needed to adequately predict activity?" To answer this, we started with the 3234-keyset, and pruned the set by random deletion of keys. It was found that a success rate of about 67% correct prediction could be maintained regardless of the selection of keys, until about 600 keys remained in the set. Below that point, classification results fell off quickly. This implies that there is a base amount of chemical structure information in the Briem structure set that must be represented, and many combinations of keys can represent this base information. The value of 600 keys is likely related to the number and diversity of structures in the data set. Including extra keys beyond this base amount did not improve the classification rate, and since the keys were included/excluded at random, they did not reduce the classification rate adversely or in a systematic manner.

It is possible that after achieving a certain rate of success, one could select additional keys that would actually reduce the overall success rate, by focusing on irrelevant structural similarities in different classes. We did not observe this in the random subset selection process, but we did see it when using another criterion for pruning the keys, the surprisal. The *surprisal* of a descriptor, a form of log likelihood ratio, is simply the negative log of the ratio of the frequency of occurrence of a given key in one activity class, divided by the frequency of occurrence in all the other classes (Figure 2). The *surprisal significance* is the surprisal divided by it's signal to noise ratio – a form of "T" statistic. As such, it can be used as a basis for selecting significant decriptors.

- Surprisal = $-\log(P_1 / P_2)$
  - probability 1 = "active" molecules
  - probability 2 = "inactive" molecules
- Assume Poisson-distributed errors
- "Surprisal Significance" =
  $\mathrm{abs}(\mathrm{Surprisal} / (S/N_{Surprisal}))$

- $S/N_{surprisal} = (1/N_a + 1/N_b)^{1/2}$

Figure 2. Definition of surprisal and surprisal significance used in this work. P_1 is the frequency of occurrence of a key in the combined active structures (for the Briem data, the five activity classes), while P_2 is the frequency of occurrence in the test set of 547 "random" structures. N_1 and N_2 are the keybit frequencies in the active and inactive classes, and S/N is the signal to noise ratio, assuming a Poisson distribution.

Using a high value of surprisal significance as a criterion for selecting keys, key subsets that were selected tended to be smaller and more efficient at predicting active structures, than the subsets obtained using random pruning. The best subsets contained about 200 keys, and showed an average prediction rate of about 71%, which is near the maximum reported by Brien and Lessel. We also examined genetic algorithms (GA) as an optimization strategy. Keyset sizes ranging from 200 to about 1600 keys were generated. The success rates ranged from 50% to 71% correct. None of the GA-generated subsets outperformed the best surprisal significance-pruned subsets.

The next question we examined was whether we could manually develop a keyset that would perform as well or better than the random, surprisal significance, and GA-pruned subsets. We took guidance form the surprisal significance pruning, the GA optimization, and the 166-keysets, along with information from publications by Bemis and Murcko that identified and classified the scaffolds and side chains found in known drug structures (3, 4). We also gave preference to the generation of series descriptors in which the counts of occurrence of atom-bond combinations varied sequentially (i.e, separate keys for 1, 2-or-more, 3-or-more, etc.) and where the bond distances in the descriptors varied sequentially (i.e., 2, 4, 6 bonds). This process eventually yielded a 324-keyset that performed as well as the best of the previously selected sets, and achieved a performance about equal to the best descriptors in the Briem paper (Figure 3). The definitions of these keys, in the form of an MDL ENKFIL file, are available from the authors.

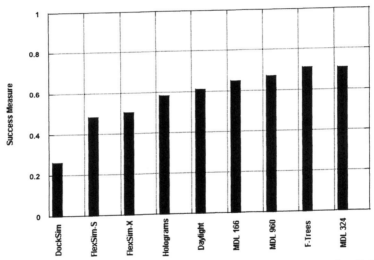

Figure 3. Performance results for the 324-key subset compared to Briem results.

Effects of Various Key Weighting Schemes

Recently there has been much interest in the prediction of ADMET properties of drugs (Absorption, Distribution, Metabolism, Excretion, and

Toxicity). This arises because many compounds that show strong binding to a receptor and high *in vitro* potency may fail as drugs, because of undesirable ADMET or pharmacokinetic properties. In our experience, attempts to use MDL keys as a basis for predicting quantitative properties such as Log P have not been successful, when decision tree and regression trees have been used (unpublished work). An alternative approach is to attempt to distinguish drug molecules from "toxic" structures using similarity-based methods.

In particular, we wanted to determine whether modifying the weights of the 166, 960, and 324-keysets in similarity calculations would improve their ability to distinguish druglike structures in the MDDR database from toxic structures in the MDL Toxicity database (17). The MDDR database we used contained 141012 unique structures (salt moieties were removed for the key calculations). The Toxicity database had 110138 structures. Only 1950 structures were present in both databases, and these were removed from the analysis. To measure success at predicting drug-likeness, we did the following:

- We selected 100 structures randomly from the combined databases (a subset of 55 MDDR structures and 45 Toxicity structures).
- Using each structure in turn as a query, we ran a similarity search, retaining all the structures with at least 50% similarity to the query structure. The weighted similarity calculations were run using the given keyset and weighting scheme. The searches retrieved a few hundred to a few thousand structures in each case.
- The percentage of structures in the result set coming from MDDR and the percent coming from Toxicity were recorded as measures of success.

We applied several different weighting schemes to determine the effect of weighting on the predictions. These included:

- Unit weighting (i.e., setting all the weights to 50).
- Weighting by 1/(Frequency of occurrence in the MDDR structures).
- Weighting by the absolute value of the MDDR/Toxicity surprisal value.

Figure 5 shows the effect of the various weighting schemes on the 166, 960, and 324-keysets. Weighting did not have a large effect on the classification results, but it is clear that the 324-keyset is superior to the others in being able to distinguish drug structures from toxic ones. Besides the accuracy of the prediction, another concern is the efficiency or selectivity of prediction – i.e., how large a result set does it take to get all the structures that are, say 50% similar to the query structure. Figure 6 shows the average result set sizes for the various weighting schemes and keysets.

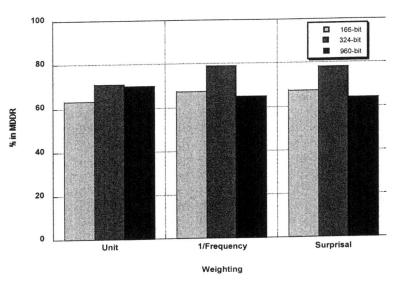

Figure 5. The effect of various weighting schemes and keysets on the prediction of "drug-likeness".

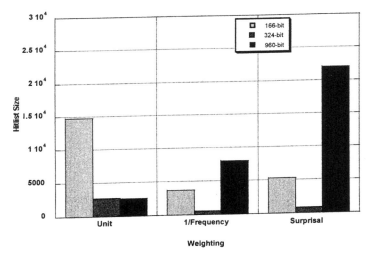

Figure 6. The effect of various weighting schemes on the hitlist size for the various keysets.

Figure 6 shows the average result set sizes for the various weighting schemes and keysets. The 324-keyset is significantly different from the other keysets, depending on the weighting that was used. In general, the 324-keyset retrieved a smaller number of structures for a given level of similarity. In practice, this would mean fewer structures to examine manually when the selection of a few specific structures is required.

Conclusions

In an extension of previous work, we have shown that substructure keysets can be optimized for predicting activity and for distinguishing drug-like from toxic structures. We examined the effects of keyset size and definition on predicting activity in a previously published data set. By redefining and selecting descriptors we were able to improve classification performance from 65% to 71% using a 324 keyset. In this work, the best criterion for pruning the keysets was found to be the surprisal significance, calculated as the ratio of the surprisal to its S/N value. We examined the effect of various weighting schemes in weighted similarity calculations to distinguish drug-like from toxic structures. The most effective weighting was observed using weights inversely proportional to the database frequency, followed by surprisal weighting. These weights yielded classification rates of nearly 80% correct. So, it is clearly possible to improve the performance of existing MDL keysets by redefining and reweighting them. The results shown here are likely very specific to the sets of structures and activities that were studied. The fact that the 324-keyset consistently performed better than any of the other keysets implies that this set may be more generally useful for drug discovery purposes.

References

1. Maggiora, G. M.; Johnson, M. A. *Concepts and Applications of Molecular Similarity; Wiley-Interscience*; Wiley-Interscience: New York, 1990; p 393.
2. Todeschini, R.; Consonni, V.; *Handbook of Molecular Descriptors; Handbook of Medicinal Chemistry*; Mannhold, R.; Kubinyi, H.; Timmerman, H., Eds.; Vol. 11; Wiley-VCH: New York, 2000.
3. Bemis, G. W.; Murcko, M. A.; The Properties of Known Drugs. 1. Molecular Frameworks. *J. Med. Chem.,* **1996**, *39*, 2887-2893.

4. Bemis, G. W.; Murcko, M. A.; The properties of Known Drugs. 2. Side Chains. *J. Med. Chem.*, **1999**, *42*, 5095-5099.
5. Nagy, M. Z.; Kuzics, S; Veszpremi, T.; Bruck, P.; Substructure Search on Very Large Files Using Tree-Structured Databases, in *Chemical Structures: The International Language of Chemistry*, Warr, W. E., Ed.; Springer-Verlag: Berlin, 1988; 127-130.
6. Willett, P. *Similarity and Clustering in Chemical Information Systems*; Wiley: New York, 1987, p 266.
7. Brown, R. D.; Martin, Y. C.; Use of Structure-Activity Data to Compare Structure-Based Clustering Methods and Descriptors for Use in Compound Selection. *J. Chem. Inf. Comput. Sci.* **1996**, *36*, 572-584.
8. McGregor, M. J.; Pallai, P. V. Clustering of Large Databases of Compounds: Using MDL "Keys" as Structural Descriptors. *J. Chem. Inf. Comput. Sci.* **1997**, *37*, 443-448.
9. Brown, R. D.; Martin, Y. C. Designing Combinatorial Library Mixtures Using a Genetic Algorithm. *J. Med. Chem.* **1997**, *40*, 2304-2313.
10. Koehler, R. T.; Villar, H. O. Design of Screening Libraries Biased for Pharmaceutical Discovery. *J. Comput. Chem.* **2000**, *21*, 1145-1152.
11. Ajay; Bemis, G. W.; Murcko, M. A. Designing Libraries with CNS Activity. *J. Med. Chem.* **1999**, *42*, 4942-4951.
12. Brown, R. D.; Martin, Y. C. The Information Content of 2D and 3D Structural Descriptors Relevant to Ligand-Receptor Binding. *J. Chem. Inf. Comput. Sci.* **1997**, *37*, 1-9.
13. Jamois, E. A.; Hassan, M.; Waldman, M. Evaluation of Reagent-Based and Product-Based Strategies in the Design of Combinatorial Library Subsets. *J. Chem. Inf. Comput. Sci.* **2000**, *40*, 63-70.
14. Briem, H.; Lessel, U. In vitro and in silico affinity fingerprints: Finding similarities beyond structural classes. *Perspec. Drug. Discov. Design.* 2000, 20, 231-244.
15. Durant, J. L.; Leland, B. A.; Henry, D. R.; Nourse, J. G. Reoptimization of MDL Keys for Use in Drug Discovery. *J. Chem. Inf. Comput. Sci.* **2002**, *42*, 1273-1280.
16. MDL Drug Data Report (MDDR). MDL, 14600 Catalina St., San Leandro, CA 94577, 2000.
17. MDL Toxicity Database, MDL, 14600 Catalina St., San Leandro, CA 94577, 2002.

Chapter 11

Clustering Compound Data: Asymmetric Clustering of Chemical Datasets

Norah E. MacCuish and John D. MacCuish

Mesa Analytics and Computing, LLC, Santa Fe, NM 87501

We investigate asymmetric clustering of compound data as a viable alternative to more commonly used algorithms in this area such as Wards, complete link, and leader algorithms. We show that the Tversky measure, more commonly applied to similarity searching in compound databases, can be used in both a hierarchical asymmetric clustering algorithm and an asymmetric variant of a popular leader algorithm as effective means to cluster 2-dimensional molecular structures for template extraction, without the size bias usually associated with more common clustering measures and methods. We show the results of the combination of these measures and algorithms with several chemical datasets.

Introduction

Cluster analysis is the study of the methods to find groups or some form of structure in data (1,2). These methods fall under the general heading of unsupervised learning. Clustering is sometimes called classification, though it is distinct from the methods, also known as classification, employed to discriminate groups that are known to be in the data a priori. Discrimination methods (3) are often used to build classification models (classifiers) for predictive modeling. The latter form of classification falls under the general rubric of supervised learning. All of these methods are within the larger study of pattern recognition (4) and multivariate statistics (5).

Clustering algorithms in turn have a complex taxonomy that is not well defined. The most general types are hierarchical (divisive and agglomerative) and partitional algorithms. A partition can be formed from a hierarchy via a level selection technique (6) such as the Davies-Bouldin (7) or the Kelley (8) heuristics. Common agglomerative hierarchical algorithms are Wards, complete link, and group average. A popular partitional or relocation method is k-means (9). Exclusion region algorithms such as Taylor-Butina (10,11) are also often used for grouping molecular structures.

The algorithms can also be divided up into forms that strictly partition the data into disjoint sets, or where cluster membership is not unique such that the sets are non-disjoint (or overlapping). Membership can also be probabilistic, where elements are assigned a probability of membership for each cluster. Fuzzy clustering (12) is an overlapping method, where membership is assigned as a grade (often between 0 and 1, but not a probability). There are also parametric methods such as mixture models. EM or Expectation Maximization is one such algorithm (13). However, these methods, with their assumption of specific distributions, tend to work best with low dimensions and they are computationally expensive.

Clustering in chemoinformatics is used for lead selection in HTS data, diversity analysis, lead hopping, compound acquisition decisions and related activities (14,15), often on large or very large data sets. Numerous clustering techniques have been employed with varying effectiveness in these pursuits (16). Algorithms must be effective in minimizing computational resources, and that the algorithm can be parallelized is often crucial with very large data sets.

Clustering compound data begins first with molecular descriptors. Such descriptors are manifold: graph-based (17), chemical properties (18), shape descriptors (19). With large data sets the speed with which to operate on molecular descriptors becomes crucial. Thus, simple binary fingerprints that encode 2D chemical structure, whether feature or path based (20,21), are very common as they are relatively easy to generate and operate on. Proximity measures for binary data are then used to compare binary molecular descriptors (22,23). These have varying properties; such as they may or may not be metric, they may be symmetric or asymmetric, or they may or may not be monotonic to one another. Thus, binary descriptors are very common, and many of the clustering techniques revolve around binary proximity measures and algorithms that can utilize them.

All of the various binary data clustering algorithms mentioned above typically use symmetric proximity measures such as the Tanimoto, Euclidean, or Ochiai measures. However, there are algorithms that can use asymmetric measures such as the Tversky measure (24,25,26). For example, there is an asymmetric, agglomerative, hierarchical, strongly connected component algorithm due to Tarjan (27), and we have transformed the Taylor-Butina algorithm mentioned above to work with asymmetric measures. Asymmetry can be used with other chemical descriptors. Examples are graph based descriptors (17) and shape descriptors (28). In addition, clusters from asymmetric algorithms need not be strictly disjoint. For instance, non-disjoint variants of hierarchical algorithms (29) have been designed, and we have created a variant of our asymmetric Taylor-Butina's algorithm to produce overlapping clusters.

Special situations arise however when using binary measures and clustering algorithms (30,31,32,33). The relative sizes of molecular structures in conjunction with certain measures can create biases (31). In addition, ties in proximity often becomes a much more serious problem with the use of binary descriptors (32,33). Ties in proximity can effect either directly or indirectly decisions within clustering algorithms, such as merging criteria in agglomerative hierarchical algorithms, or partitioning decisions. Algorithms in turn may include fundamental or implementation decisions that result in an ambiguous clustering.

We show how asymmetric methods are largely equivalent to the common and popular clustering methods in use in chemoinformatics.

Motivation

Anecdotal evidence from the chemical information industry suggests that the Tversky asymmetric measure is used with considerable efficacy in similarity searching -- where, given one compound, a database is searched for similar compounds. It measures, via its parameterization of similarity, to what extent is a single molecular structure either super- or sub-similar to others. This gives rise to two possibly different proximities, hence the asymmetry. Similarity searching uses one-to-many comparisons in one direction, whereas clustering typically uses many-to-many symmetric comparisons.

Asymmetric clustering algorithms in other fields have not been tried to our knowledge in chemoinformatics. Our research suggests that using asymmetric measures and asymmetric clustering algorithms may yield important new methods that provide insight into template extraction or determining substructures in common. Our original interest in asymmetry was in the hopes that it would help avoid the serious shortcomings of ties in proximity when using binary measures and various clustering algorithms. This benefit is marginal at best, but does not obviate the other benefits of the use of asymmetry. More generally, asymmetric clustering algorithms can be used with non-binary descriptors as well, such as clustering shape descriptors or graph-based descriptors.

Asymmetry

Symmetric measures such at the Tanimoto and Euclidean measures of proximity are simply binary relations such that the proximity between two molecular structures is a single value. With asymmetric measures however, the proximity now has possibly two values. The proximity between structure A and structure B is not the same as the proximity of structure B and structure A.

Tversky Measure

The Tversky measure is a parameterization of the Tanimoto measure. The parameters allow one to treat the measures as asymmetric. The Tanimoto and Tversky measures are defined in Equation 1 and Equation 2 for comparisons of molecular structures represented by binary bit strings.

$$\text{Tanimoto} = c / [a + b + c] \qquad (1)$$

$$\text{Tversky} = c / [\alpha a + \beta b + c] \qquad (2)$$

a = unique bits set in molecular structure A
b = unique bits set in molecular structure B
c = common bits set in structures A and B

Optional constrain: $0 <= \alpha <= 1$, and $\beta = 1 - \alpha$

Given the optional constraint above, if $\alpha = 1$, then the Tversky value will approach 1 as the number of unique bits set in molecular structure A tend to 0. So if A is a substructure of B in terms of the descriptor, with $\alpha=1$ and $\beta=0$, then the Tversky measure is maximized to 1. The choice of α enables the Tversky value to determine how closely B contains A's structure given the descriptor. Tversky will give a different value when the question is asked in the opposite direction. When $\beta = 1$ the question becomes, does B fit into A? The unique bits of B will determine the Tversky result for the comparison. Strictly speaking, α and β can be set independently and outside of the 0 to 1 interval. However, even with the optional constraint, shades of superimposition can be obtained by setting α and β between 0 and 1. Thus, the Tversky measure is an asymmetric measure as it provides two different values depending on the values of the parameters.

Asymmetric Clustering Algorithms

An algorithm due to Tarjan recursively calculates the strongly connected components of the directed graph formed by the asymmetry of the proximity measure used. This algorithm is computationally equivalent to computing the minimum spanning tree of a graph. Practical algorithms exist for this problem with bounds of $O(E \log \log V)$, where E is the number of edges in the graph and V is the number of vertices. If N is the number of structures this bound is $O(N^2 \log \log N)$. Generating the proximity matrix with a given threshold can decrease the number of edges substantially.

We have created an asymmetric variant of the Taylor-Butina clustering algorithm. Though this algorithm is effectively $O(N^2)$, we can generate a threshold matrix such that the algorithm can operate on a sparse matrix, improving both the space and time requirements. The asymmetric clustering algorithms is as follows:
1. Create threshold graph.
2. Find true singletons: all those compounds with both zero in and out degree.
3. Find the structure with the largest in or out degree. This becomes a group and is excluded from subsequent consideration. (The structure is known as the *representative structure.*)
4. Repeat 3 until no compounds exist with positive in or out degree.
5. Optional: Assign remaining compounds, false singletons, to the group that contains the nearest neighbor in terms of either in or out degree.

In and out degree refers to the edges of the directed graph – in this case the edges formed by the Tversky asymmetry.

A Simple Experiment

Our task is to show that asymmetric algorithms are at least as effective as the commonly used algorithms. We therefore assembled a small simple set with three basic classes of compounds with some intermixing. The set contains HIV ligands with known antiviral activity, in three structural classifications, Azido Pyrmidines (NCI identification number labeled with a P), Benzodiazepines (labeled with a B), and Pyrimidine Nucleosides (labeled with a N). Binary Fingerprints of length 166 were generated with MDL keys.

For comparison we used Wards, complete link, Taylor-Butina and two asymmetric algorithms, Tarjan's and an asymmetric version of Taylor-Butina. With Wards and complete link we use 1-Tanimoto, or the Soergel, dissimilarity measure, as ties in proximity are less likely to impact the resulting hierarchies over the use of the Euclidean measure. For the two asymmetric algorithms we used the Tversky measure.

Results

In Figure 1, we display the results for a complete link clustering of the dataset with a Tanimoto dissimilarity cutoff of 0.15. At this cut off, Figure 1 displays the groupings found, in which nearly all compounds are clustered into one of the three classes in the dataset. At this cut off, seven groups noted by Roman numeral labels are found. In Figure 2, the results for Wards clustering with a merging criterion of 0.2, are displayed. The same structural classes are grouped together as well as the same seven groups (labeled by Roman Numerals) are found. In Figure 3, the results from the asymmetric hierarchical clustering algorithm of Tarjan are displayed. At Tversky dissimilarity of 0.09 merging criteria, the Tarjan algorithm is able to also generate groupings that are delineated by the compound classes. The groups formed differ in one compound NCI number 635034 is moved from group V into group IV and an outlier NCI number 620753 is included in group V. (Note, in groups I, II, and III, there are three compounds merged at the same level. This is an artifact of the strongly connected component merging criterion rather than each triple having the same similarity to one another when merged. The merging level is determined by the last similarity value that completes the strongly connected component.)

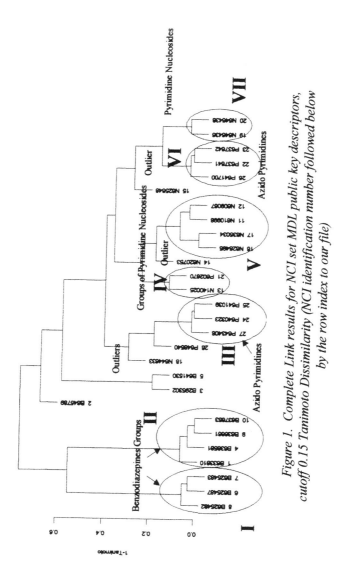

Figure 1. Complete Link results for NCI set MDL public key descriptors, cutoff 0.15 Tanimoto Dissimilarity (NCI identification number followed below by the row index to our file)

164

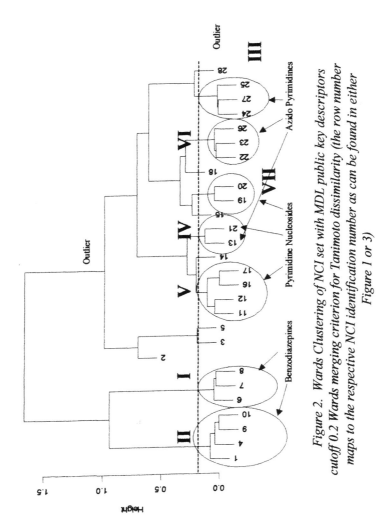

Figure 2. Wards Clustering of NCI set with MDL public key descriptors cutoff 0.2 Wards merging criterion for Tanimoto dissimilarity (the row number maps to the respective NCI identification number as can be found in either Figure 1 or 3)

165

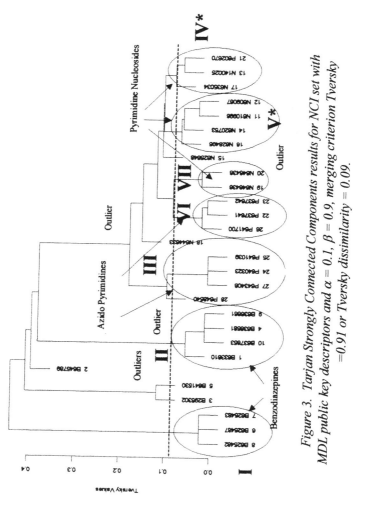

Figure 3. Tarjan Strongly Connected Components results for NCI set with MDL public key descriptors and $\alpha = 0.1$, $\beta = 0.9$, merging criterion Tversky $= 0.91$ or Tversky dissimilarity $= 0.09$.

In Table I, the results for symmetric Taylor-Butina at a similarity threshold cut off of 0.85 are given and in Table II the asymmetric Taylor-Butina clustering results are displayed for an $\alpha = 0.1$, $\beta = 0.9$ and a Tversky similarity threshold of 0.91. In the case of the symmetric grouping algorithm, groups I, II, the Benzodiazepines are the same as in Figure1 for the complete link results, and the group III Azido Pyrimidines are also pulled out of the set by this symmetric algorithm. The Group IV and V Pyrimidine Nucleosides are merged together into one group that also contains an outlier NCI 620753. The remaining Pyrimidine Nucleoside group, VII is merged into one group with Group VI, the remaining Azido Pyrimidine group.

In Table II we note a reduction in the number of singletons to two. Two of the singletons in Figure 1, NCI 295302 and NCI 641530 are merged into a third group of Benzodiazepines. The asymmetric grouping algorithm also finds the two Benzodiazepine groups I and II, as noted in Figure 1. Like the symmetric results the Pyrimidine Nucleosides and the Azido Pyrimidines are merged into three groups instead of five. Group VI and VII are merged together into one group, just like in the symmetric case. This grouping also contains the outlier NCI 625648 (15). Group V occurs as in Figure 1 with one nearby outlier included NCI 620753 and one group member moved to the final group, NCI 610998. The final group is a merging of Groups III and IV including the outlier mentioned previously as well as NCI 648540. The asymmetric algorithm results contain fewer singletons and no false singletons. Both results create groups with similar structural classes, mixing occurs for each slightly differently.

Conclusions

Asymmetric clustering provides a new approach to clustering methods which have become commonplace in the pharmaceutical industry. Two degrees of freedom in similarity decisions opens the door to novel strategies for grouping and classification of chemical datasets. The application of these techniques will judge their utility and scalability in facilitating problem solving in areas such as high throughput screening results, compound acquisition decisions, and diversity assessments.

A thorough investigation of the issue of ties in proximity and clustering ambiguity in general, with both symmetric and asymmetric measures and clustering algorithms is a future topic being investigated.

Table 1. Taylor-Butina Symmetric at Tanimoto Similarity Threshold of 0.85

True Singletons	Benzodiazepines	Pyrimidine Nucleosides and	Azido Pyrimidines
	II; I	*IV+V+outlier*	*VII + VI;III*
B645789 (2)	B633810 (1)	N610998 (11)	N646438 (20)
B295302 (3)	B638581 (4)	N609067 (12)	P637641 (22)
B641530 (5)	B636661 (9)	N140025 (13)	P637642 (23)
N625648 (15)	B637653 (10)	N620753 (14)	P641700 (26)
N644633 (18)		N628495 (16)	N646436 (19)
P648540 (28)	B625487 (6)	N635034 (17)	
	B625483 (7)	P602670 (21)	P640323 (24)
	B625482 (8)		P641039 (25)
			P643408 (27)

NOTE: False Singletons N646436 (19) and P602670 (21). Integers in parentheses allow comparison with Figure 2. Roman numerals correspond to group labels in Figure 1.

Table II. Asymmetric Taylor-Butina Tversky Similarity Threshold at 0.91

True Singletons	Benzodiazepines	Pyrimidine Nucleosides and	Azido Pyrimidines
	II; I; VIII	*V+outlier;VI+VII+outlier*	*III+IV+outlier*
B645789 (2)	B633810 (1)	N609067 (12)	N610998 (11)
N644633 (18)	B638581 (4)	N620753 (14)	N140025 (13)
	B636661 (9)	N628495 (16)	P602670 (21)
	B637653 (10)	N635034 (17)	P640323 (24)
			P641039 (25)
	B625487 (6)	N625648 (15)	P641700 (26)
	B625483 (7)	N646436 (19)	P643408 (27)
	B625482 (8)	N646438 (20)	P648540 (28)
		P637641 (22)	
	B295302 (3)	P637642 (23)	
	B641530 (5)		

NOTE: No False Singletons, integer numbers in parentheses allow for comparison with Figure 2. Roman numerals correspond to group labels in Figure 1.

References

1. *Algorithms for Clustering Data*; Jain, A. K.; Dubes, R. C. Prentice Hall Advanced Reference Series: Englewood Cliffs, NJ, 1988.
2. *Finding Groups in Data: An Introduction to Cluster Analysis*; Kaufman, L.; Rousseeuw, P. J. John Wiley & Sons, Inc: New York, NY, 1990.
3. *Discriminant Analysis and Statistical Pattern Recognition*; MacLachlan, G.J. John Wiley & Sons, Inc: New York, NY, 1992.
4. *Pattern Classification*; Duda, R. O.; Hart, P. E.; Stork, D. G. 2nd edn. John Wiley & Sons, Inc., New York, NY, 2001.
5. *Applied Multivariate Statistical Analysis*; Wichern, D. W.; Johnson, R. A. Prentice Hall: Englewood Cliffs, NJ, 2002.
6. Wild, D.J.; Blankley, C.J. Comparison of 2D Fingerprint Types and Hierachy Level Selection Methods for Structural Grouping Using Ward's Clustering, *J. Chem. Inf. Comput. Sci.* **2000**, *40*, 155-162.
7. Davies, D. L.; Bouldin, D. W. A cluster separation measure. *IEEE Transactions on Pattern Analysis and Machine Intelligence PAMI, 1,* **1979**, 224-227.
8. Kelley, L. A.; Gardner, S. P.; Sutcliffe, M. J. An automated approach for clustering an ensemble of NMR-derived protein structures into conformationally-related subfamilies. *Protein Eng.* **1996**, *9*, 1063-1065.
9. Faber, V. Clustering and the Continuous *k*-Means Algorithm. *Los Alamos Science*, **1994**, 22.
10. Taylor, R. Simulation Analysis of Experimental Design Strategies for Screening Random Compounds as Potential New Drugs and Agrochemicals, *J. Chem. Inf. Comput. Sci.* **1995**, *35*, 59-67.
11. Butina, D. Unsupervised Data Base Clustering Based on Daylight's Fingerprint and Tanimoto Similarity: A Fast and Automated Way To Cluster Small and Large Data Sets, *J. Chem. Inf. Comput. Sci.* **1999**, *39*, 747-750.
12. Beni, G.; Liu, X. A Least Biased Fuzzy Clustering Method *IEEE Transactions on Pattern Analysis and Machine Intelligence* **1994**, 16(9),954-960.
13. *The EM Algorithm and Extensions*; McLachlan, G. J.; Krishnan, T. John Wiley & Sons: New York, NY, 1997.
14. Warr, W. A. Combinatorial Chemistry and Molecular Diversity: An Overview, *J.Chem. Inf. Comput. Sci.* **1997**, 27,134-140.
15. Brown, R.D.; Martin, Y.C. Use of Structure-Activity Data To Compare Structure-Based Clustering Methods and Descriptors for Use in Compound Selection, *J.Chem.Inf. Comput. Sci.* **1996**, 36, 572-584.

16. Downs, G. M.; Barnard, J. M. Clustering Methods and Their Uses in Computational Chemistry, *Reviews in Computational Chemistry*; Vol. 18, Lipkowitz, K. B. and Boyd, D. B., Eds; Wiley-VCH: New York, **2002,** 1-40.
17. Raymond, J. W.; Gardiner, E. J.; Willett, P. RASCAL: Calculation of Graph Similarity using Maximum Common Edge Subgraphs. *J. Chem. Inf. Comput. Sci.* **2002** *45* (6), 631-644.
18. eduSoft, LC., Richmond, Va. Home page: http://www.edusoft-lc.com/
19. Putta, S.; Lemmen, C.; Beroza, P.; Greene, J. A Novel Shape—Feature Based Approach to Virtual Library Screening, *J. Chem. Inf. Comput. Sci* **2002**, *42*, 1230-1240.
20. MDL Information Systems, Inc., San Leandro, CA. Home page: http://www.mdli.com/.
21. Daylight Chemical Information Systems, Inc., Mission Viejo, CA. Home page:]http://www.daylight.com/.
22. Rhodes, N.; Willett, P. Bit-String Methods for Selective Compound Acquisition. *J. Chem. Inf. Comput. Sci* **2000**, 40, 210-214.
23. Willet, P.; Barnard, J. M.; Downs, G. M. Chemical Similarity Searching. *J.Chem. Inf. Comput. Sci*. **1998**, *38* (6), 983.
24. Tversky, A. *Psychological Reviews,* **1977,** *84(4),* 327-352.
25. Hubalek, Z. Coefficients of Association and Similarity, Based on Binary (Presence – Absence) Data: An Evaluation. *Biol. Rev.* **1982,** 57, 669-689.
26. Bradshaw, J. Introduction to the Tversky similarity measure. *MUG '97 – 11th Annual Daylight User Group Meeting*, February, 1997.
27. Tarjan, R. An Improved Algorithm for Hierarchical Clustering Using Strong Components. *Inf. Process. Lett.* (*IPL*) **1983**, *17*, 37-41.
28. MacCuish, N. E.; MacCuish, J. D. Shape Clustering with Tversky Similarity. Manuscript in Preparation.
29. Nicolaou, C. A.; MacCuish, J. D.; Tamura, S. Y. A new multidomain clustering algorithm for lead discovery that exploits ties in proximities. In *Rational Approaches to Drug Design*; Proceedings of the 13th European Symposium on Quantitative Structure—Activity Relationships. Dusseldorf, Aug 27—Sept 1 2000; Prous Scientific.
30. Flower, D. R. On the Properties of Bit String-Based Measures of Chemical Similarity. *J. Chem. Inf. Comput. Sci.* **1998** *38*, 379-386.
31. Dixon, S. L.; Koehler, R. T. The Hidden Component of Size in Two Dimensional Fragment Descriptors: Side Effects on Sampling in Bioactive Libraries. *J. Med. Chem.* **1999**, *42*, 2887-2900.

32. Godden, J. W.; Xue L.; Bajorath, J. Combinatorial Preferences Affect Molecular Similarity/Diversity Calculations Using Binary Fingerprints and Tanimoto Coefficients. *J. Chem. Inf. Comput. Sci.* **2000**, *40*, 163-166.
33. MacCuish, J.; Nicolaou, C.; MacCuish, N. Ties in Proximity and Clustering Compounds. *J. Chem. Inf. Comput. Sci.* **2001** *41* (1), 143-146.

Chapter 12

From Decision Tree to Heterogeneous Decision Forest: A Novel Chemometrics Approach for Structure–Activity Relationship Modeling

Weida Tong[1,*], Huixiao Hong[2], Hong Fang[2], Qian Xie[2], Roger Perkins[2], and John D. Walker[3]

[1]Center for Toxicoinformatics, Division of Biometry and Risk Assessment, National Center for Toxicological Research, 3900 NCTR Road, HFT 20, Jefferson AR 72079
[2]Northrop Grumman Information Technology, Jefferson, AR 72079
[3]TSCA Interagency Testing Committee (ITC), U.S. Environmental Protection Agency (7401), Washington, DC 20460
*Corresponding author: telephone: (870) 543–7142; fax: (870) 543–7662; email: wtong@nctr.fda.gov

> The techniques of combining the predictions of multiple classification models to produce a single model have been investigated for many years. In earlier applications, the multiple models to be combined have been developed by altering the training set. The use of these so-called resampling techniques, however, enhance the risk of reducing predictivity of the models to be combined and/or over fitting the noise in the data, which might result in poorer prediction of the composite model than the individual models. In this paper, we suggest a novel approach, named Heterogenious Decision Forest (HDF), that combines multiple Decision Tree models. Each Decision Tree model is developed using a unique set of descriptors. When models of similar predictive quality are combined using the HDF method, quality compared to the individual models is consistently and significantly improved in both training and testing steps. An example will be presented for prediction of binding affinity of 232 chemicals to the estrogen receptor.

U.S. government work. Published 2005 American Chemical Society.

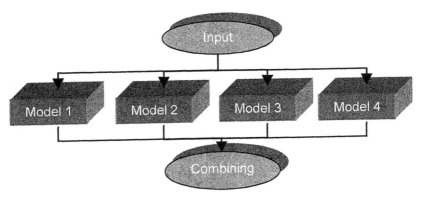

Figure 1. A schematic presentation of combining the results of four models.

Combining (or ensemble) forecast is a statistical technique that combines the results of multiple individual models to reach a single prediction[1]. The overall scheme of the technique is shown in Figure 1, where the individual models are normally developed using an Artificial Neural Network (ANN)[2-4] or Decision Tree[5-6]. A thorough review of this subject can be found in a number of papers[7-9].

In most cases, individual models are developed using a portion of chemicals randomly selected from the original dataset[10]. For example, a dataset can be randomly divided into two sets, 2/3 for training and 1/3 for testing. A model developed with the training set will be accepted if it gives satisfactory predictions for the testing set. A set of predictive models is generated by repeating this procedure, and the predictions of these models are then combined when predicting a new chemical. The training set can also be generated using more robust statistical "resampling" approaches, such as Bagging[11] or Boosting[12].

Bagging is a "bootstrap" ensemble method by which each model is developed on a training set that is generated by randomly selecting chemicals from the original dataset[11]. In the selection process, some chemicals may be repeated more than once while others may be left out so that the training set is the same size as the original dataset. In Boosting, the training set for each model also is the same size as the original dataset. However, each training set is determined based on the performance of the earlier model(s); chemicals that are incorrectly predicted by the previous model are chosen more often than chemicals that were correctly predicted in the next training set[12]. Boosting, Bagging and other resampling approaches have all been reported to improve predictive accuracy.

The resampling approaches use only a portion of the dataset for constructing the individual models. Since each chemical in a dataset encodes some Stucture

Activity Relationship (SAR) information, reducing the number of chemicals in a training set for model construction will weaken most individual models' predictive accuracies. It follows that reducing the number of chemicals also reduces the improvement in a combining system gained by the resampling approach. Moreover, Freund and Schapire reported that some resampling techniques could be at risk of overfitting the noise in the data, which leads to much worse prediction from multiple models[12].

The idea of combining multiple models implicitly assumes that one could not identify all aspects of the underlying variable relationship, and thus different models are able to capture it for prediction. Combining several identical models produces no gain. The benefit of combining multiple models can be realized only if individual models give different predictions. An ideal combined system should consist of several accurate models that disagree as much as possible.

In this paper, a novel combining forecast approach is explored that classifies a new chemical by combining the predictions from multiple decision tree models. This method is named Heterogenious Decision Forest (HDF). A HDF model consists of a set of individually trained decision trees that are developed using unique sets of descriptors. Our results suggest that the HDF model is consistently superior to any individual trees that are combined to produce the forest in both training and validation steps.

Materials and Methods

Heterogenious Decision Forest (HDF) Algorithm

The important aspects of the HDF approach were: 1) each individual model in Figure 1 was developed using a *distinct* set of descriptors that was explicitly excluded from all other models, thus ensuring each individual model's unique contribution to making predictions; and 2) the quality of all models in HDF is *comparable* to ensure that each model significantly contributes to the prediction. The development of the HDF algorithm consists of the following steps:
1. The algorithm can be initiated with either a pre-defined number of models (N) to be combined or a misclassification threshold to set a quality criterion for individual models. The former case is illustrated in this paper.
2. A tree is constructed without pruning. The tree identifies the minimum number of misclassified chemicals (*MIS*) for a given dataset. *MIS* then serves as a quality criterion to guide individual tree construction and pruning in the following iterative steps 3-6.
3. A tree is constructed and pruned. The extent of pruning is determined by the *MIS*. The pruned tree assigns a probability (0-1) to each chemical in the dataset.

4. The descriptors used in the previous model are removed from the original descriptor pool, and the remaining descriptors are used for the next tree development.
5. Steps 3 and 4 are repeated until no additional model with misclassifications ≤ *MIS* can be developed from the unused portion of the original pool of descriptors.
6. If the total number of models is less than *N*, the *MIS* is increased by 1, and the steps 3 to 5 are repeated. Otherwise, multiple decisions from individual trees are combined using a linear combination method, where the mean value of the probabilities for all trees is used to determine the classification of a chemical. A chemical with the mean probability larger than 0.5 is designated as active while a chemical with a mean value less than 0.5 is designated as inactive.

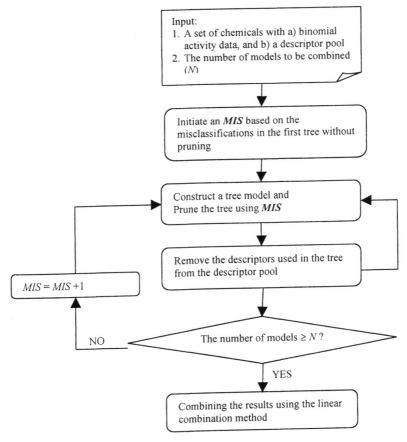

Figure 2. Flowchart of the HDF algorithm.

Tree development

The development of a tree model consists of two steps, tree construction and tree pruning. In the tree construction process, a parent population is split into two children nodes that become parent populations for further splits. The splits are selected to maximally distinguish the response descriptors in the left and right nodes. Splitting continues until chemicals in each node are either in one activity category or can not be split further to improve the model. To avoid overfitting the training data, the tree needs to be cut down to a desired size using tree cost-complexity pruning. In the present application, the method for the tree development is described by Clark and Pregibon[13] as implemented in S-Plus, which is a variant of the Classification and Regression Tree (CART) algorithm. It employs deviance as the splitting criterion. The HDF algorithm is written in S language and run in S-Plus software.

Model assessment

Misclassification and concordance are used to measure model quality. Misclassification is the number of chemicals misclassified in a tree model, while concordance is the number of correct predictions divided by the total number of predictions.

NCTR dataset

A large and diverse estrogen dataset, called the NCTR dataset[14-15], was used in this study. The NCTR dataset contains 232 structurally diverse chemicals[16], of which 131 chemicals were found to actively bind to an estrogen receptor[17] while 101 are inactive[18] in a competitive estrogen receptor binding assay.

Descriptors

More than 250 descriptors for each molecule were generated using Cerius 2 software (Accelrys Inc., San Diego, CA 92121). These descriptors were categorized as 1) conformational, 2) electronic, 3) information content, 4) quantum mechanical, 5) shape related, 6) spatial, 7) thermodynamic, and 8) topological. The descriptors were preprocessed by removing those with no variance across the chemicals. A total of 197 descriptors were used for the final study.

Results

Figure 3 gives a plot of misclassification versus the number of combined decision trees. The number of misclassifications varies inversely with the number of trees. The reduction in misclassification is greatest in the first four trees, where more than ½ the misclassifications were eliminated. A forest comprising seven distinct trees eliminated about 2/3 of the misclassifications of the initial decision tree.

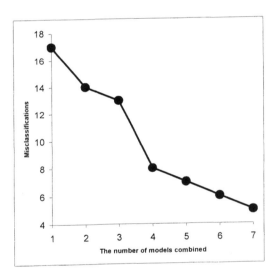

Figure 3. Relationship of misclassifications with the number of trees combined in HDF.

Table 1 provides more detailed results on the models of the HDF and the decision trees combined. Based on misclassifications, all HDF models (different combinations) perform better than any individual decision tree. Of 202 original descriptors, 88 were ultimately used for the forest combining seven decision trees. The progressive decrease in misclassifications as decision trees are successively added to the forest demonstrates how each unique descriptor set contributes uniquely to the aggregate predictive ability of the forest. Generally, decision trees with fewer "branches" are expected to perform better because the descriptors are better able to encode the functional dependence of activity on structure. Table 1 also shows the expected trends of both more descriptors and more branches in the decision trees as the descriptors better able to encode the activity versus structure dependency are successively removed from the descriptor pool.

Table 1. The results of 7 individual trees and their combination performance

Tree ID	Number of Descriptors used	Number of branches	Misclassifications in	
			Each Tree	Combination
1	10	13	17	17
2	10	13	19	14
3	12	15	17	13
4	12	14	17	8
5	15	18	19	7
6	16	19	20	6
7	13	17	18	5

Table 2 gives a comparison of decision tree and HDF as measured by chemicals predicted as active that are actually inactive (false positives) and chemicals predicted as inactive that are actually active (false negatives). The decision tree being compared corresponds to that in the first row of Table 1 that has 17 misclassifications. The forest being compared in Table 2 corresponds to the bottom row in Table 1 where seven decision trees are combined and for which there are five misclassifications. In the Table 2 comparison, the decision tree utilizes 10 descriptors and produces nine false negatives and eight false positives. In contrast, the forest model utilizes 88 unique descriptors and produces four false negatives and one false positive, a marked improvement in the prediction performance compared to the decision tree.

Table 2. Comparison of model performance between Decision Tree and HDF

Experiment Results	Decision Tree Prediction		HDF Prediction	
	A*	I*	A	I
A = 131	122	9	127	4
I = 101	8	93	1	100

* A = Active; I = Inactive

Among the many schemes to combine multiple decision trees, we evaluated linear combination and voting. The voting method uses the majority of votes to classify a chemical. The linear combination method uses the mean of probabilities of the individual decision trees. We found the two methods to produce the same results (results not shown), and chose linear combination because a tie vote cause problem in the voting method.

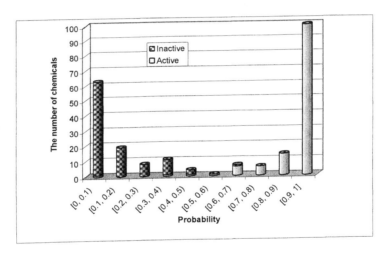

Figure 4. Distribution of active/inactive chemicals across the probability bins in HDF. The probability of each chemical was the mean value calculated over all individual trees in HDF. A chemical with probability larger than 0.5 was designated as active while less than 0.5 was inactive.

HDF assigns a mean probability of the combined trees using the linear combination approach. Figure 4 shows the concordance results of the HDF prediction of the NCTR dataset in ten even intervals between 0 and 1. Analysis shows that the interval 0.7 – 1.0 has an average concordance of 100% of true positives, and the interval 0.0 – 0.3 has an average concordance of 98.9% true negatives. The vast majority of misclassifications occur in the 0.3 – 0.7 probability range where the average concordance is 78%.

A more robust validation of the predictive performance was conducted by dividing the NCTR data set into a training component comprising two-thirds, or 155, of the chemicals and a testing component comprising the remaining 77 chemicals. Both HDF and Decision tree models were constructed for a random selection of the training set, and then used to predict the testing set. This was repeated 2000 times to give the concordance results shown in Figure 5. Figure 5

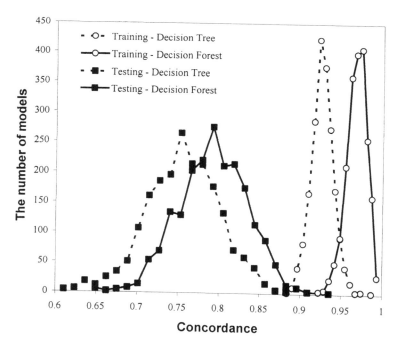

Figure 5. Comparison of the model results between Decision Tree and HDF in a validation process. In this method, the dataset was divided into two groups, 2/3 for training and 1/3 for testing. The process was repeated 2000 times. The solid line is associated with the results from HDF while the dash line is for Decision Tree. The quality of a model in both training (circle) and predication (square) was assessed using concordance that was calculated by dividing the misclassifications by the number of training chemicals in the training step and by the number of testing chemicals in prediction, respectively. The position of a dot (circle or square) on the graph identifies the number of models with a certain value of concordance.

gives on the Y-axis the number of times out of 2000 that a model attained the concordance value given on the X-axis. The consistently better predictive average concordance of the HDF is readily discernable, as is the narrower distribution for prediction of the training set versus the test set. Both leave-one-out and leave-10-out validation tests were also performed and showed a similar trend (results not shown).

Discussion

We presented a novel combining forecast approach, HDF, that combines predictions of individually trained Decision Trees, each developed using a set of unique descriptors. The method was illustrated by classifying 232 chemicals into estrogen and non-estrogen receptor-binding categories. We demonstrated that DHF yielded better classification and prediction than Decision Tree in both training and validation steps.

A SAR equation can be generalized as Bio = $f(D_1, D_2, ..., D_n)$, where Bio is biological activity data (binomial data in classification) and D_1 to D_n are descriptors. This equation implies that the variance in Bio is explained in a chemistry space defined by the descriptors ($D_1 ... D_n$). Accordingly, HDF can be understood as a pooling result of SAR models that predict activity within their unique chemistry spaces. Since each SAR model is developed using a unique set of descriptors, the difference in their prediction is maximized. Thus, it is safe to assume that combining multiple valid SAR models that use unique sets of descriptors into a single decision function should provide better estimation of activity than that separately predicted by the individual models.

A number of commercial software packages, including CODESSA (Semichem, Shawnee, KS), Cerius2 (Accelrys Inc., San Diego, CA) and Molconn-Z (eduSoft, LC, Richmond, VA), enables a large volume of descriptors to be generated for SAR studies. HDF takes advantage of this large volume of descriptors by aggregating the information of structural dependence on activity represented from each unique set of descriptors. Unlike the re-sampling techniques used in most combining forecast approaches, all training chemicals are included in each decision tree to be combined in the HDF, thus maximizing the SAR information.

It is important to note that there is always a certain degree of noise associated with biological data, and particularly the data generated from a High

Throughput Screen process. Thus, optimizing SAR models inherently risks over fitting the noise, a result most often observed using ANNs. Since the combination scheme of HDF is not a fitting process, some noise introduced by individual SAR models will be cancelled when combining predictions. Moreover, using Decision Tree to construct HDF offers additional benefits because the quality of a tree can be adjusted in the pruning process using the *MIS* parameter as a figure of merit for model quality. The *MIS* parameter is an indicator of noise, enabling the modeler a way to reduce over fitting of the noise.

HDF can be used for priority setting in both drug discovery and regulatory applications. The objective of priority setting is to rank order from most important to least important a large number of chemicals for experimental evaluation. The purpose of some priority setting in drug discovery is to identify a few lead chemicals, but not necessarily all potential ones. In other words, relatively high false negatives are tolerable, but false positives need to be low. In the example we presented, chemicals predicted to be active with probability > 0.7 were shown to have 100% concordance with experimental data, thus demonstrating its use for lead selection.

In contrast, a good priority setting method for regulatory application should generate a small fraction of false negatives. False negatives constitute a crucial error, because they will receive a relatively lower priority for experimental evaluation. In the example we presented, chemicals predicted to be inactive with probability < 0.3 were shown to have 98.9% concordance with experimental data, thus demonstrating its use for regulatory application.

Acknowledgement

The research is funded under the Inter-Agency Agreement between the U.S. Environmental Protection Agency and the U.S. Food and Drug Administration's National Center for Toxicological Research. The authors also gratefully acknowledge the American Chemistry Council and the FDA's Office of Women's Health for partial financial support.

References

1. Bates, J. M. and Granger, C. W. J. The combination of forecasts. *Oper. Res. Quart.* **1969**, *20*, 451-468.
2. Opitz, D. and Shavlik, J. Actively searching for an effective neural-network ensemble. *Connect. Sci.* **1996**, *8*, 337-353.

3. Krogh, A. and Vedelsby, J. *Neural network ensembles, cross validation and active learning*; Tesauro, G.; Touretzky, D. and Leen, T.: MIT Press, 1995; 7, 231-238.
4. Maclin, R. and Shavlik, J. Combining the predictions of multiple classifiers: Using competitive learning to initialize neural networks. *Proc. 14th Int. Joint Conf. Intel.* **1995**, 524-530.
5. Drucker, H.; Cortes, C.; Jackel, L.D.; LeCun, Y.; Vapnik, V. *Boosting and other ensemble methods. Advanced in Neural Information Processing Systems 8*; Touretsky, D.S.; Mozer, M.C.; Hasselmo, M.E.; Morgan Kaufmann: San Francisco **1996**, 479-485
6. Quinlan, J. Bagging, boosting and c4.5. *Proc. 13th Nat. Conf. Artif. Intel.* **1996**, 725-730.
7. Bunn, D. W. *Expert use of forecasts: Bootstrapping and linear models*; Wright, G. and Ayton, P.: Wiley, **1987**; 229-241.
8. Bunn, D. W. Combining forecasts. *Eur. J. Operat. Res.* **1988**, *33*, 223-229.
9. Clemen, R. T. Combining forecasts: A review and annotated bibliography. *Int. J. Forecast.* **1989**, *5*, 559-583.
10. Maclin, R. and Opitz, D. An empirical evaluation of Bagging and Boosting. *Proc. 14th Nat. Conf. Artif. Intel.* **1997**, 546-551.
11. Breiman, L. Bagging predictors. *Mach. Learn.* **1996**, *24*, 123-140.
12. Freund, Y. and Schapire, R. Experiments with a new Boosting algorithm. *Proc. 13th Int. Conf. Mach. Learn.* **1996**, 148-156.
13. Clark, L. A. and Pregibon, D. *Tree-based models*; Chambers and Hastie, **1997**; Chapter 9, 413-430.
14. Blair, R.; Fang, H.; Branham, W. S.; Hass, B.; Dial, S. L.; Moland, C. L.; Tong, W.; Shi, L.; Perkins, R. and Sheehan, D. M. Estrogen receptor relative binding affinities of 188 natural and xenochemicals: Structural diversity of ligands. *Toxicol. Sci.* **2000**, *54*, 138-153.
15. Branham, W. S.; Dial, S. L.; Moland, C. L.; Hass, B.; Blair, R.; Fang, H.; Shi, L.; Tong, W.; Perkins, R. and Sheehan, D. M. Binding of phytoestrogens and mycoestrogens to the rat uterine estrogen receptor. *J. Nutrit.* **2002**, *132*, 658-664.
16. Fang, H.; Tong, W.; Shi, L.; Blair, R.; Perkins, R.; Branham, W. S.; Dial, S. L.; Moland, C. L. and Sheehan, D. M. Structure activity relationship for a large diverse set of natural, synthetic and environmental chemicals. *Chem. Res. Toxicol.* **2001**, *14*, 280-294.
17. Shi, L. M.; Tong, W.; Fang, H.; Perkins, R.; Wu, J.; Tu, M.; Blair, R.; Branham, W.; Walker, J.; Waller, C. and Sheehan, D. An integrated "4-

Phase" approach for setting endocrine disruption screening priorities - Phase I and II predictions of estrogen receptor binding affinity. *SAR/QSAR Environ. Res.* **2002**, *13*, 69-88.

18. Hong, H.; Tong, W.; Fang, H.; Shi, L. M.; Xie, Q.; Wu, J.; Perkins, R.; Walker, J.; Branham, W. and Sheehan, D. Prediction of estrogen receptor binding for 58,000 chemicals using an integrated system of a tree-based model with structural alerts. *Environ. Health Persp.* **2002**, *110*, 29-36.

Indexes

Author Index

Bajorath, Jürgen, 41
Bennett, Kristin P., 111
Berglund, Anders, 31
Bi, Jinbo, 111
Breneman, Curt M., 111, 127
Brown, Steven D., 15
Cramer, Steven, 111
Davidson, Charles E., 127
Durant, Joseph L., Jr., 145
Fang, Hong, 173
Feudale, Robert, 15
Garg, Rajni, 97
Henry, Douglas R., 145
Holliday, John D., 77
Hong, Huixiao, 173
Katt, William, 127
Lavine, Barry K., 1, 127

MacCuish, John D., 157
MacCuish, Norah E., 157
Milne, G. W. A., 55
Perkins, Roger, 173
Pettersson, Fredrik, 31
Salim, Naomie, 77
Song, Minghu, 111
Stahura, Florence L., 41
Sukumar, N., 111
Tan, HuWei, 15
Tong, Weida, 173
Tugcu, N., 111
Walker, John D., 173
Willett, Peter, 77
Workman, Jerome, Jr., 1
Xie, Qian, 173
Xue, Ling, 41

Subject Index

A

Acetone, calibration and prediction errors, 26t
Acute toxicity LD_{50}, equation, 70
ADMET (adsorption, distribution, metabolism, excretion, and toxicity), property prediction, 152–153
AIDS dataset
  similarity property principle, 88, 90f, 91f
  See also Chemical similarity and dissimilarity
Algorithms
  clustering, 158
  genetic, in polymer design, 68
  heterogenious decision forest (HDF), 175–176
  Morgan, for structure searching, 59–60
  See also Genetic algorithms (GA)
Angor treatment dose (ATD), equation, 70
Anti-HIV-1 protease drugs
  comparative quantitative structure-activity relationship (QSAR) studies, 103–107
  See also Cheminformatics
Artificial neural networks (ANN)
  prediction model, 174
  quantitative structure-retention relationship (QSRR), 112
Asymmetry
  clustering compound data, 160–162
  Tversky measure, 160–161
Azido pyrimidines
  asymmetric Taylor–Butina Tversky similarity cutoff of 0.91, 166, 168t
  Taylor–Butina symmetric at Tanimoto similarity cutoff of 0.85, 166, 167t

B

Bagging
  bootstrap ensemble method, 174
  prediction results, 119–120, 123
  See also Quantitative structure-retention relationship (QSRR)
Barnard Chemical Information (BCI), bit-string, 82
Baseline correction methods, calibration and prediction errors, 26t
Benzodiazepines
  asymmetric Taylor–Butina Tversky similarity cutoff of 0.91, 166, 168t
  Taylor–Butina symmetric at Tanimoto similarity cutoff of 0.85, 166, 167t
Binary data, clustering algorithms, 159
Bioinformatics
  chemometrics application, 12
  distinguishing between, and chemoinformatics, 49–52
Biological activity
  predictions for drug discovery, 146
  See also MDL substructure search keys
Bioseparation, ion exchange chromatography (IEC), 112
Bit-strings
  Barnard Chemical Information (BCI), 82
  datasets, 82
  Daylight, 82

generation of fragment, 80, 82
randomization approach, 82
similarity search against target using, -based Tanimoto coefficient, 81*f*
Tanimoto coefficient, 78
Tanimoto frequency distributions of intermolecular similarity values, 84*f*
UNITY, 82
See also Chemical similarity and dissimilarity
Blood
classification type, 33
See also Microarrays
Bone marrow
classification type, 33
See also Microarrays
Boosting
pattern recognition analysis, 135
resampling approach, 174
Bootstrap aggregation, bagging, 116

C

Calibration
acetone, 26*t*
full, 18
robust, 18–19
robust modeling through multi-scale, 27–29
strategy for flawed, 18
See also Multivariate calibrations
Calibration model transfer, instrumental standardization, 16–17
Cargill. *See* Corn samples
Chemical Abstracts Service, structure searching, 59
Chemical graph theory
approximations, 62
chemical mathematics, 57–58
Chemical mathematics
acute toxicity LD_{50}, 70
angor treatment dose (ATD), 70

approximations in chemical graph theory, 62
Connectivity index, 66
Dendral Project, 58–59
DNA computing, 71–72
genetic algorithms, 68
graph theory, 57–58
Hamiltonian path problem, 71–72
Hosoya Index, 65–66
ID Number index, 66–67
isomer enumeration, 62–63
Kappa Indices, 67
modern programs and compute power, 60–61
Morgan algorithm, 59–60
non-deterministic polynomial (NP) time problem, 71–72
reverse processing, 67
statistics, 69–70
structure searching, 58–61
substructure and superstructure searching, 61
topological index (TI), 69
topological indices, 63–67
traveling salesman problem, 71–72
Wiener Index, 64–65
Chemical similarity and dissimilarity
analysis of coefficients, 83, 86–87
analysis of similarities, 82–83
applicability of similar property principle, 87–89
comparison of Tanimoto, Euclidean, and Cosine frequency distributions of intermolecular similarity values, 85*f*
Cosine coefficient, 86–87
dissimilarity-based compound selection (DBCS), 79
Euclidean complement coefficient, 86
frequency distributions of Tanimoto similarity values, 84*f*
generation of fragment bit-strings, 80, 82

low-valued Tanimoto coefficients, 79–80
measuring structural similarity, 78–79
plots for AIDS database, 90f, 91f
plots for ID Alert database, 92f, 93f
similarity search against target using bit string based Tanimoto coefficient, 81f
Tanimoto coefficient, 78–79, 86
Chemical structure space, dimensionality, 146
Cheminformatics
bio-chemical Bio-database, 99
calculated partition coefficient (ClogP) in octanol/water, 102
ClogP values of FDA approved HIV protease drugs, 106
comparative QSAR studies in anti-HIV-1 protease drugs, 103–107
CQSAR database (QSAR=quantitative structure-activity relationships), 99
CQSAR program and database, 99–101
definition, 98
electronic parameters, 102–103
hydrophobic parameter, 102
hypothesis in design of new analogs, 98
MERLIN feature of CQSAR database, 101
molecular descriptors, 101–103
outliers in QSAR analysis, 107
physicochemical parameters, 101–103
physico-chemical Phys database, 99
quantitative structure-activity relationships (QSAR), 98
role in drug design, 103–107
search for new lead in drug-design, 101
searching CQSAR database, 100–101
steric parameters, 103
structure of CQSAR database, 100
substituent selection in molecular design, 101
tasks, 98
Chemistry
macroscopic and microscopic, 56–57
mathematical, 57
place among the sciences, 56
See also Chemical mathematics
Chemoinformatics
boundaries between molecular modeling, 45–46
boundaries to computational disciplines, 42
challenges, 44–46
chemometrics application, 11
clustering, 158
converging disciplines, 49–52
current trends, 43–44
distinguishing between bio- and, 49, 50f
distinguishing between drugs and non–drugs, 43
driving force and popylarity, 42
drug discovery, 42–43, 44, 49–52
education, 46
examples of popular topics, 43
hierarchical organization of topics in bio- and, 51f
interfaces with experimental research, 47–49
iterative screening, 48f
opportunities, 46–52
outlook, 52
pharmaceutical industry, 43
predicting binding energies and affinities, 45
predicting in vivo properties of clinical candidates, 45
schematic illustrating reduction of biological and chemical data, 50f

similar algorithms, diverse
  applications, 50–52
 summary, 52
 target validation chemistry, 49
 term, 41
 unresolved scientific questions, 45–46
 virtual (VS) and high-throughput screening (HTS), 47–49
Chemometrics
 applications, 11–12
 chemoinformatics, 11
 collinearity, 4, 9
 definitions, 2
 global modeling, 19–20
 image analysis, 11
 implications of data, 2–3
 indirect observation approach, 2–3
 microarrays, 11–12
 molecular dynamic (MD) simulations, 12
 multivariate analysis methods, 4, 6f, 7f
 octane number variation, 3–4, 5f
 partial least squares (PLS), 10
 principal component analysis (PCA), 4, 8–10
 redundancy, 4, 9
 sensor performance, 11
 soft modeling in latent variables, 3–4
 summary of paradigm for learning, 3
 term, 2
 variance, 8–9
 See also Structure activity relationship (SAR)
Classification, clustering, 158
Cluster analysis, unsupervised learning, 158
Clustering algorithms
 asymmetric, 160, 161–162
 asymmetric hierarchical, of Tarjan, 162, 165f
 binary data, 159
 hierarchical and partitional, 158
Clustering compound data
 algorithms, 158
 asymmetric clustering algorithms, 161–162
 asymmetric hierarchical clustering algorithm of Tarjan, 162, 165f
 asymmetric Taylor–Butina Tversky similarity cutoff of 0.91, 166, 168t
 asymmetric variant of Taylor–Butina clustering algorithm, 161
 asymmetry, 160–162
 binary data clustering algorithms, 159
 chemoinformatics, 158
 dataset with Tanimoto dissimilarity cutoff of 0.15, 162, 163f
 molecular descriptors, 158–159
 motivation, 159–160
 symmetric Taylor–Butina at similarity cutoff of 0.85, 166, 167t
 Tversky measure, 160–161
 Wards clustering with merging criterion of 0.2, 162, 164f
Collinearity
 principal component analysis (PCA), 4, 9
 soft modeling, 4
Computational disciplines, chemoinformatics research, 42
Computer analysis
 musk odorants and structurally related nonmusk compounds, 128
 See also Odor structure relationships (OSR)
Connectivity index, chemical mathematics, 66
Corn samples
 piecewise orthogonal signal correction (POSC) and OSC, 22, 23f
 starch prediction, 29f

Cosine coefficient
  analysis of coefficients, 83, 86–87
  analysis of similarities, 82–83
  comparison of coefficients, 85f
  See also Chemical similarity and dissimilarity
CQSAR (quantitative structure-activity relationship)
  program and database, 99–101
  See also Cheminformatics
Cross-validation, acetone calibration, 26t

D

Database. *See* Cheminformatics
Daylight Chemical Information Systems
  bit-string, 82
  design of substructure search and similarity keys, 147–148
Decision tree
  combining, for heterogenious decision forest (HDF), 174–175
  performance of, versus HDF, 178–179
  prediction model, 174
  tree development, 177
  See also Structure activity relationship (SAR)
Dendral Project, structure searching, 58–59
Deoxyribonucleic acid (DNA) computing, chemical mathematics, 71–72
Descriptor calculation program, DRAGON, 146
Descriptor generation
  definition of descriptors from support vector regression (SVR) feature selection, 120t
  MOE program, 115–116, 123
  reconstruction program, 114–115
  transferable atom equivalents (TAEs), 114–116
Dimensionality, chemical property space, 146
Disease classification
  gene expression, 32
  See also Microarrays
Dissimilarity coefficient
  complement of Tanimoto coefficient, 78
  See also Chemical similarity and dissimilarity
DNA. *See* Deoxyribonucleic acid (DNA) computing
DRAGON program, descriptor calculation, 146
Drug discovery
  challenges of chemoinformatics, 44–46
  chemoinformatics, 42–43
  heterogenious decision forest (HDF), 183
  ion exchange chromatography (IEC), 112
  pharmaceutical company marketing and, 98
  predicting biological activity, 146
  virtual screening (VS), 112
  See also Cheminformatics; Chemoinformatics; Quantitative structure–retention relationship (QSRR)
Drug structures. *See* MDL substructure search keys
Dual-domain calibration, definition, 28
Dual-domain regression
  advantages, 29
  partial least squares (PLS) model, 28–29
  two-step procedure, 27

E

Education, chemoinformatics, 46
Electron-density derived descriptors, transferable atom equivalents (TAEs), 114–115, 123, 129
Electronic parameters, CQSAR program, 102–103
Electronic surface properties, transferable atom equivalents (TAEs), 115t
Enumeration of isomers, chemical mathematics, 62–63
Euclidean complement coefficient
  AIDS and ID Alert databases, 88, 91f, 93f
  analysis of coefficients, 83, 86–87
  analysis of similarities, 82–83
  comparison of coefficients, 85f
  definition, 86
  See also Chemical similarity and dissimilarity

F

Feature selection, classical support vector regression (SVR) variation, 116–118
Fingerprints, topological descriptor, 146
Fitness function, pattern recognition analysis, 133–134
Flawed calibration, strategy for, 18

G

Gasolines
  Mallat wavelet prism transform, 25f
  variation in spectra, 3–4, 5f
  wavelet transformation of spectral data, 24f
Gene expression
  microarray technologies, 31–32
  See also Microarrays
Genetic algorithms (GA)
  block diagram for pattern recognition, 136f
  chemical mathematics, 68
  pattern recognition, 132–135
  See also Odor structure relationships (OSR)
Global modeling, robust calibration, 19–20
Graph theory
  approximations, 62
  chemical mathematics, 57–58

H

Hamiltonian path problem, chemical mathematics, 71–72
Hammett electronic parameters, CQSAR program, 102–103
Heterogenious decision forest (HDF)
  algorithm, 175–176
  flowchart of algorithm, 176f
  pooling of structure-activity relationship (SAR) models, 182
  priority setting method, 183
  software packages, 182
  See also Structure activity relationship (SAR)
High-throughput screening (HTS), chemoinformatics, 47–49
HIV-1 protease inhibitors
  comparative molecular field analysis (CoMFA), 105
  FDA approved, 106
  mortality and morbidity of AIDS, 103
  See also Cheminformatics
Hosoya Index, chemical mathematics, 65–66
Hydrophobic parameter, CQSAR program, 102

I

ID Alert dataset
  similarity property principle, 88, 92f, 93f
  See also Chemical similarity and dissimilarity
ID number index, chemical mathematics, 66–67
Image analysis, chemometrics application, 11
Implementation strategy, classical support vector regression (SVR) variation, 116
Indirect observation, chemometrics, 2–3
Ion exchange chromatography (IEC) bioseparation, 112
  protein retention dataset, 114
Isomer enumeration, chemical mathematics, 62–63
Iterative screening, chemoinformatics, 47–49

J

Jaccard coefficient, Tanimoto coefficient, 78
Jarvis–Patrick clustering method, nearest neighbor identification, 79

K

Kappa Indices, chemical mathematics, 67
Keysets
  constructing better, 150–155
  definition and selection of keys, 150
  key definitions in MDL, 149t
  MDL databases, 148
  performance, 152
  predicting activity, 150–151
  surprisal significance, 151
  See also MDL substructure search keys

L

Latent variables, soft modeling, 3–4
Leukemia
  molecular classification of samples, 39
  See also Microarrays
Linear free energy relationship (LFER), approach using descriptors, 98
Linear least squares, technique averaging data, 10

M

Macroscopic chemistry, microscopic and, 56–57
Mathematical chemistry, 57
Mathematics. *See* Chemical mathematics
MDL Drug Data Report (MDDR) database
  computing similarity of compounds, 150
  distinguishing druglike substances, 153
MDL substructure search keys
  ADMET (absorption, distribution, metabolism, excretion, and toxicity) property prediction, 152–153
  defining MDL keys, 147–149
  definition and selection of keys, 150
  descriptor calculation program DRAGON, 146
  dimensionality of chemical structure space, 146

effect of weighting schemes and keysets on prediction of drug-likeness, 154f
effect of weighting schemes on hitlist size for keysets, 154f
effects of various key weighting schemes, 152–155
ENKFIL file or table to define substructure key, 148
fingerprints, 146
genetic algorithms (GA) as optimization, 151–152
key definitions for first 24 keys in MDL 166-keyset, 149t
new use for substructure keys, 146–147
number of keys for predicting activity, 150–151
performance rates, 152
predicting biological activity from structure, 146
substructure "keys", 146
success rates, 151–152
surprisal of descriptor, 151
surprisal significance, 151
Tanimoto coefficient, 147
Measurements, chemometrics, 2–3
Methylcyclopropane
graph theory, 57–58
Morgan numbering, 59–60
Microarrays
ALL (lymphoid origin) or AML (myeloid origin) bone marrow or blood samples, 33
chemometrics application, 11–12
comparing w1 and regression coefficients, 37–38
disease classification, 32
experimental, 33
finding cutoff value, 34–35, 38–39
gene expression, 31–32
generation of ranking list, 33–34
methods, 33–35
molecular classification of leukemia samples, 39

partial least squares (PLS) risks, 32–33
PLS-DA model, 33
PLS-DA model for ALL/AML, 35, 36f
score plot for PLS-DA model discriminating ALL and AML samples, 36f
tool in functional genomics, 32
validation of prediction (Q2) and correlation (R2Y), 35, 37
Microscopic chemistry, macroscopic and, 56–57
Misclassifications
model assessment, 177
plot of, versus number of combined decision trees, 178f
See also Structure-activity relationship (SAR)
Model improvement
localized preprocessing with wavelets, 22, 24–27
projection methods for, 20–22
Modeling
boundaries between molecular, 45–46
soft, in latent variables, 3–4
Modeling method, partial least squares (PLS), 10
MOE program, descriptor generation, 115–116, 123
Molecular descriptors
clustering compound data, 158–159
CQSAR program, 101–103
Molecular dynamic (MD) simulations, chemometrics application, 12
Morgan algorithm, structure searching, 59–60
Multivariate analysis methods, soft modeling in latent variables, 4, 6f, 7f
Multivariate calibrations
calibration model transfer, 16–17
dual-domain calibration, 27–28

dual-domain partial least squares
(DDPLS) regression, 29
full calibration, 18
global modeling, 19–20
improving calibration model, 16
limitation, 16
local analysis for robust calibration
model creation, 17–20
local preprocessing with wavelets
for model improvement, 22, 24–
27
orthogonal signal correction (OSC),
20–21
performance of POSC and OSC on
representative near infrared
(NIR) data, 22, 23f
piecewise OSC (POSC)
preprocessing, 21
projection methods for improving
model performance, 20–22
robust calibration, 18–19
robust modeling through multi-
scale calibration, 27–29
standardization of instrumental
aspects, 16–17
starch prediction of Cargill corn
samples, 29f
strategy for flawed calibration, 18
tool, 16
Multivariate projection method,
classification of leukemia samples,
39
Musk compounds
discriminant analysis, 128–129
musk data set, 130–131
structural class, 131f
See also Odor structure
relationships (OSR)

N

Near infrared (NIR) spectra
acetone calibration and prediction
errors, 26t

multivariate calibration, 16
orthogonal signal correction (OSC)
and piecewise OSC (POSC), 22,
23f
Noise, fitting with biological data,
182–183
Noise variation, principal component
analysis (PCA), 9
Non-deterministic polynomial (NP)
time problem, chemical
mathematics, 71–72
Nonmusk compounds
discriminant analysis, 128–
129
musk data set, 130–131
structural class, 131f
See also Odor structure
relationships (OSR)

O

Octane number, variation in spectra,
3–4, 5f
Odor structure relationships (OSR)
applying principal component
analysis (PCA) to training data
set, 136, 138f
block diagram of genetic algorithm
(GA) for pattern recognition,
136f
boosting, 135
class and sample weight equations,
133
computer analysis using descriptors
and pattern recognition, 128
discriminant analysis to
differentiate musk and nonmusk,
128–129
electron density derived
descriptors, 130–131
facilitating design of new odorants,
129
fitness function of pattern
recognition, 133–134

genetic algorithm (GA) for pattern recognition, 132–133
limited success of past efforts, 129–130
molecular descriptors by pattern recognition GA, 137, 140t
musk data set, 130–131
pattern recognition analysis, 132–135
pattern recognition GA, 130
principal component plot of 331 compounds and 16 TAE derived descriptors, 137, 139f
principal mapping experiment for 331 compounds and 871 descriptors, 136, 138f
property encoded surface translator (PEST) algorithm, 132
representative musk and nonmusk compound classes, 131f
sample hit count, 133–134
transferable atom equivalent (TAE) descriptor methodology, 129
Olfaction, phenomenon, 128
Orthogonal signal correction (OSC)
piecewise OSC (POSC) preprocessing, 21
POSC and OSC performance, 22, 23f
principle, 20

P

Paradigm of learning, chemometrics, 3
Partial least squares (PLS)
dual-domain regression, 28–29
modeling method, 10
quantitative structure-retention relationship (QSRR), 112
risks for microarray data, 32–33
Pattern recognition analysis
block diagram of genetic algorithm (GA) for, 136f
class and sample weight equations, 133
fitness function, 133–134
genetic algorithm, 129–130
genetic algorithm (GA), 132–133
property encoded surface translator (PEST) algorithm, 132
sample hit count (SHC), 133–134
See also Odor structure relationships (OSR)
Pharmaceutical industry chemoinformatics, 43
See also Chemoinformatics
Physicochemical parameters, CQSAR program, 101–103
Piecewise orthogonal signal correction (POSC)
performance for corn samples, 22, 23f
preprocessing, 21
Piecewise preprocessing, orthogonal signal correction (OSC), 21
Polymer design, genetic algorithms, 68
Prediction errors, acetone calibration, 26t
Preprocessing for robust models, information loss, 27
Principal component analysis (PCA)
axes defining measurement space, 8f
collinearity, 4, 9
redundancy, 4, 9
variance, 8–10
See also Odor structure relationships (OSR)
Principal component regression (PCR), quantitative structure-retention relationship (QSRR), 112
Priority setting method, heterogenious decision forest (HDF), 183
Property encoded surface translator (PEST) algorithm
pattern recognition analysis, 132

surface property hybrid descriptors, 130, 131, 132
Protein retention dataset
  ion exchange chromatography (IEC) system, 114
  predictions, 119–120, 121*f*
Pyrimidine nucleosides
  asymmetric Taylor–Butina Tversky similarity cutoff of 0.91, 166, 168*t*
  Taylor–Butina symmetric at Tanimoto similarity cutoff of 0.85, 166, 167*t*

Q

Quantitative structure-activity relationship (QSAR)
  dimensionality of chemical property space, 146
  paradigm, 98
  statistical mechanics, 69
  See also Cheminformatics
Quantitative structure-property relationship (QSPR), statistical mechanics, 69
Quantitative structure-retention relationship (QSRR)
  bagging (Bootstrap Aggregation), 116
  classical support vector regression (SVR) variation (ν-SVR), 116
  dataset generation, 114–116
  definition of descriptors from SVR feature selection, 120*t*
  descriptor generation, 114–116
  electron-density derived descriptors, 114–115, 123
  feature selection, 116–118
  general framework of feature selection scheme, 118*f*
  margin, 117
  nonlinear regression bagging models, 118–119
  prediction scatter plot using non-linear SVR model, 121*f*
  protein retention dataset, 114
  star plots generation process, 122*f*
  statistical algorithms, 112
  support vector regression (SVR), 112–114
  SVR feature selection and bagging prediction results, 119–120, 123
  transferable atom equivalents (TAEs), 114–115, 123

R

Reconstruction program, descriptor generation, 114–115
Redundancy
  principal component analysis (PCA), 4, 9
  soft modeling, 4
Regulatory applications, heterogenious decision forest (HDF), 183
Resampling approaches
  bagging, 174
  boosting, 174
  structure-activity relationship (SAR), 174–175
Reverse processing, chemical mathematics, 67
Robust calibration
  global modeling, 19–20
  isolating key wavelengths in near infrared (NIR), 19
  local analysis, 17–20
  See also Multivariate calibrations
Robust modeling, multi-scale calibration, 27–29
Root mean squared error of cross-validation (RMSECV), calculation for corn samples, 22, 23*f*
Root mean squared error of prediction (RMSEP), calculation, 22

S

Sensor performance, chemometrics application, 11
Signal variation, principal component analysis (PCA), 9
Similarity
  applicability of similar property principle, 87–89
  measurement of structural, 78
  See also Chemical similarity and dissimilarity
Simulations, chemometrics application, 12
Single-domain regression models, combination of set, 27–28
Soergel coefficient, complement of Tanimoto coefficient, 78
Soft modeling, latent variables, 3–4
Software packages, heterogenious decision forest (HDF), 182
Standardization, instrumental aspects of calibration, 16–17
Starch prediction, corn samples, 29*f*
Star plot
  generation process, 122*f*
  support vector regression (SVR) model, 120
Statistics, chemical mathematics, 69–70
Steric parameters, CQSAR program, 103
Structural similarity, measurement, 78
Structure-activity relationship (SAR)
  cause and effect relation, 98
  combining forecast, 174, 182
  combining multiple models, 175
  commercial software packages, 182
  comparison of model performance between decision tree and heterogenious decision forest (HDF), 179, 180*t*
  comparison of model results between decision tree and HDF in validation process, 181*f*
  degree of noise associated with biological data, 182–183
  descriptors, 177–178
  distribution of active/inactive chemicals across probability bins in HDF, 180*f*
  equation, 182
  flowchart of HDF algorithm, 176*f*
  HDF algorithm, 175–176, 182
  linear combination method, 180
  materials and methods, 175–178
  model assessment, 177
  National Center for Toxicological Research (NCTR) dataset, 177
  performance of individual decision tree vs. HDF, 178–179
  plot of misclassification vs. number of combined decision trees, 178*f*
  priority setting method, 183
  probability of combined trees using linear combination approach, 180–181
  resampling approaches, 174–175
  schematic of combining model results, 174*f*
  tree development, 177
  validation, 181–182
  voting method, 180
  See also Cheminformatics
Structure/property-activity, methodologies, 42
Structure searching
  chemical mathematics, 58–61
  Dendral Project, 58–59
  Morgan algorithm, 59–60
  substructure, 61
  superstructure, 61
Substructure, structure searching, 61
Substructure keys
  Tanimoto coefficient, 146–147
  topological descriptor, 146
  typical entry in ENKFIL table, 148*f*
Substructure search. *See* MDL substructure search keys
Superstructure, structure searching, 61

Support vector machine (SVM) regression
 modeling approach, 112, 124
 See also Quantitative structure-retention relationship (QSRR)
Support vector regression (SVR)
 characteristics, 112–114
 classical SVR variation, 116
 definition of descriptors from feature selection, 120*t*
 ε-insensitive losses as training error, 113
 feature selection and bagging prediction, 119–120, 123
 graphical depiction of ε-insensitive loss function, 113*f*
 prediction scatter plot using non-linear SVR model, 121*f*
 regularization factor, 113
 regularization parameter, 113
 star plots generation process, 122*f*
 See also Quantitative structure-retention relationship (QSRR)
Surprisal, descriptor, 151
Surprisal significance, form of T statistic, 151

T

Tanimoto coefficient
 analysis of coefficients, 83, 86–87
 analysis of similarities, 82–83, 84*f*
 binary form, 78–79
 comparison of coefficients, 85*f*
 definition, 78, 147
 distribution, 79
 Jarvis–Patrick clustering method, 79
 low-valued, 79–80
 similarity search against target using bit-string based, 81*f*
 substructure keys, 146–147
 See also Chemical similarity and dissimilarity

Tanimoto dissimiliarity cutoff, dataset clustering, 162, 163*f*
Tanimoto measure, equation, 160
Target validation chemistry, chemoinformatics, 49
Tarjan, clustering algorithm, 162, 165*f*
Taylor–Butina clustering algorithm
 asymmetric, Tversky similarity cutoff, 166, 168*t*
 asymmetric variant, 161–162
 symmetric, 166, 167*t*
Term
 chemoinformatics, 41
 chemometrics, 2
 robust calibration, 18–19
Topological index (TI), statistical mechanics, 69
Topological indices
 chemical mathematics, 63–67
 Connectivity index, 66
 Hosoya Index, 65–66
 ID Number index, 66–67
 Wiener Index, 64–65
Toxicity. *See* MDL substructure search keys
Transferable atom equivalents (TAEs)
 electron-density derived descriptors, 114–115, 123, 129
 See also Odor structure relationships (OSR)
Traveling salesman problem, chemical mathematics, 71–72
Tversky measure
 asymmetry, 160–161
 equation, 160

U

UNITY, bit-string, 82

V

Vapor pressure, calculation, 69

Variance, principal component analysis (PCA), 8–10
Variation, spectra of gasolines and octane number, 3–4, 5f
Virtual screening (VS)
  chemoinformatics, 47–49
  drug design, 112
  See also Quantitative structure-retention relationship (QSRR)

W

Wards clustering, merging criterion, 162, 164f
Wavelet preprocessing
  acetone from near infrared (NIR) spectra, 26–27
  discrete wavelet transform (DWT), 24
  gasoline spectral data, 24f
  localized, for model improvement, 22, 24–27
  Mallat wavelet prism transform, 25f
Weight
  effects of weighting schemes, 152–155
  substructure keys, 147
Wiener Index, chemical mathematics, 64–65
Wold, Herman, partial least squares (PLS), 10

Printed in the USA/Agawam, MA
June 8, 2012

566441.063